ADVANCES IN PROBABILITY
and Related Topics

Advances in Probability
and Related Topics

VOLUME 1

Edited by Peter Ney

DEPARTMENT OF MATHEMATICS
UNIVERSITY OF WISCONSIN
MADISON, WISCONSIN

MARCEL DEKKER, INC., New York 1971

MARCEL DEKKER, INC.

95 Madison Avenue, New York, New York 10016

LIBRARY OF CONGRESS CATALOG CARD NUMBER 75-79066

ISBN No.: 0-8247-1491-1

Introduction to the Series

This is the first of a series of volumes on current work in *probability and related areas.* Each volume will contain about four to six articles of a research and/or expository nature. Authors will have greater flexibility in terms of the length of papers and amounts of expository material they wish to include than they would have in the standard mathematics journals; they will thereby be free to present a more complete development of their subject. Considerable emphasis will be placed on clarity of exposition. Although the central theme of the series is probability theory, the term 'related topics' will be liberally interpreted (e.g. ergodic theory, harmonic analysis), and articles detailing developments in applications of probability will also be encouraged.

In the present volume, inasmuch as the several chapters deal with independent topics, a common index was felt to be inappropriate and was therefore not included.

Authors wishing to submit papers for inclusion in the series are invited to send their manuscripts to the editor.

P.N.

Contributors to Volume 1

HARRY FURSTENBERG, Hebrew University, Jerusalem, Israel

L. H. HARPER, The Rockefeller University, New York, New York

I. I. HIRSCHMANN, Jr., Department of Mathematics, Washington University, St. Louis, Missouri

J. P. KAHANE, Université de Paris, Institut Henri Poincaré, Paris, France

GIAN-CARLO ROTA, Massachusetts Institute of Technology, Cambridge, Massachusetts

Contents

Contents of Volume 2

ADVANCES IN PROBABILITY
and Related Topics

Random walks and discrete subgroups of Lie groups*

by Harry Furstenberg

HEBREW UNIVERSITY
JERUSALEM, ISRAEL

* *This research was supported in part by the Air Force Office of Scientific Research.*

Probabilists have been somewhat partial to group theory – at least to group structures – since the beginning of their subject. The classical limit theorems deal with combining large numbers of random variables in accordance with a group law – addition. Markov chains have been defined for arbitrary state spaces, but the theory is most striking for the classical random walk on the m-dimensional lattice – the free abelian group with m generators ([9], [22]). Diffusions may be defined on general topological spaces, but the one with the richest theory is Brownian motion, the diffusion on Euclidean space invariant under the group of rigid motions. More recently, random walks have been studied on more general groups ([7], [16]), and steps have been taken to extend classical limit theorems to a wider class of groups ([12], [13], [14]).

In what follows we shall attempt to repay a part of our debt to group theory by proving a purely algebraic theorem using probabilistic methods. As far as we know, this type of result has not been obtained any other way.*

1. Statement of the Problem

Let G be a non-compact connected Lie group. (From now on all Lie groups are connected.) A discrete subgroup $\Gamma \subset G$ is *full* if the quotient space G/Γ carries finite invariant measure. This condition means that there is a subset D in G with finite left-invariant Haar measure such that the translates $D\gamma$, $\gamma \in \Gamma$ cover G. (In other words, Γ has a 'fundamental domain' with finite measure.) When this is the case we will also say that G is the *envelope* of Γ. We shall also refer

* Added in proof: Since this was written, an alternative proof to our main theorem has been given by S. P. Wang (to appear) based on work by Kazhdan: "Connection of the dual space of a group with the structure of its closed subgroups", *Funktsional'nyi Analizi Ego Prilozheniya*, **1**, 71–74 (1967).

to G as an envelope of Γ in case Γ is merely isomorphic to a full subgroup of G.

There are numerous examples of full subgroups of Lie groups. We mention some of the simplest. An m-dimensional lattice is full in the Euclidean space $\boldsymbol{R}^m$. In this case the quotient space is actually compact (i.e., there exists a compact fundamental domain). When this happens, we say the subgroup is *cocompact*. An example of a full subgroup that is not cocompact is the group of $m \times m$ integer unimodular matrices, $SL(m, \boldsymbol{Z})$, as a subgroup of the corresponding group with real entries, $SL(m, \boldsymbol{R})$. The finiteness of the measure in this case follows from a general theorem of this type given in [2]. Naturally, if G/Γ carries finite measure and $[\Gamma: \Gamma']$ is finite for a subgroup Γ', then G/Γ' carries finite measure. Thus the subgroups of finite index of $SL(m, \boldsymbol{Z})$ – for example, the congruence subgroups: $\Gamma_l = \{\gamma \in SL(m, \boldsymbol{Z}): \gamma \equiv \text{identity} \pmod{l}\}$ – are full subgroups of $SL(m, \boldsymbol{R})$.

Another large class of examples is obtained as follows. If Ω is a compact orientable 2-dimensional manifold, it may be given the structure of a Riemann surface. It must have as simply connected covering space either the sphere, the full plane or the half-plane (= unit disc). If the genus (number of handles) of Ω is greater than 1, then the covering space is the upper half-plane. The group of covering transformations of Ω is then a discrete subgroup Γ of the set of directly conformal transformations of the upper half-plane. The latter is the so-called hyperbolic group, $G = SL(2, R)/\{\pm 1\}$. Since Ω is compact, G envelopes Γ.

A number of questions arise naturally at this point.

(i) Which discrete groups possess envelopes altogether?

(ii) If an envelope exists, to what extent is it unique?

(iii) Can we go so far as to say that if $\Gamma_1 \cong \Gamma_2$, with G_1 and G_2 enveloping Γ_1 and Γ_2 respectively, then the isomorphism of Γ_1 and Γ_2 extends to an isomorphism of G_1 and G_2?

If the discrete group is a finitely generated nilpotent group, then the answer to all three questions is affirmative ([3], [20], [18]). In the general case much less can be said. It is known that the statement in (iii) is false for certain groups. The counterexample comes about by considering two homeomorphic, but conformally inequivalent, compact Riemann surfaces. The isomorphism of the fundamental groups does not extend to an automorphism of the hyperbolic group.

In connection with question (ii), it is clear that we cannot expect uniqueness in a strict sense, since the product of an envelope with a compact group will again be an envelope. However it makes sense to ask whether the envelope is unique modulo compact components. In particular we may ask whether it is possible for two Lie groups G_1 and G_2, neither of which possesses a compact component (i.e., G_i cannot be written $G' \times K$ with K compact), to envelope the same discrete group. It is to this question that we shall make a modest contribution. Namely, we shall show that there are at least some pairs of distinct semi-simple Lie groups which cannot both envelope the same discrete group. More specifically, what can be shown is that the unimodular groups $SL(r, \boldsymbol{R})$ for $r \neq 2$ cannot envelope a group that is enveloped by $SL(2, \boldsymbol{R})$. While this result is not very far-reaching, the method of proof promises to yield more information of this type.

Some consequences of this result have nothing to do with Lie groups. For example, we deduce from it that subgroups of finite index of $SL(2, \boldsymbol{Z})$ cannot occur with finite index in $SL(r, \boldsymbol{Z})$ for $r \neq 2$. In particular, the free group with 2 generators which occurs with index 6 in $SL(2, \boldsymbol{Z})$ ([21]) cannot occur with finite index in $SL(r, \boldsymbol{Z})$ if $r \neq 2$.

Question (i), that of the existence of an envelope, is more recondite and we shall have nothing to contribute to its solution. However, if, instead, we ask whether a group possesses a semi-simple envelope then some simple necessary conditions are known. One such condition relates to probabilistic considerations (it follows partially from Kesten's work [16]), and we state it here without proof. Let Γ be finitely generated with generators $g_1, g_2, \ldots, g_r$. Let H_n denote the number of distinct elements that can be expressed by products of no more than n of the g_i and their inverses. A necessary condition for Γ to be a full subgroup of a semi-simple Lie group is that $\lim_{n+\infty}, (\log H_n)/n > 0$, i.e., that H_n grow exponentially with n. Note that this condition is independent of the choice of generators of Γ. Abelian groups clearly violate this condition, whereas the free groups with finitely many generators satisfy it.

2. Random Walks on Groups and Boundaries

We turn now to question (ii) and discuss informally our approach to the problem. The question of isomorphism of envelopes is essentially that of finding out how much information regarding an envelope is determined by the discrete group Γ. The answer we give, very loosely stated, is that Γ and its envelope must have the same behavior 'at infinity'. The problem, of course, will be to make this statement precise. Now the same idea may be expressed in terms of 'boundaries'. Suppose it were possible to attach, in a canonical way, a boundary to every locally compact group which, in some sense, represents the set of all limits of group elements converging to ∞. We expect that the set of limits would be the same for a group as for a full subgroup (if we are optimistic). It would then follow that a given discrete group Γ can have as envelope only such Lie groups whose 'boundary' coincides with that of Γ.

The actual picture is much more complicated. Nonetheless, we may give an example where this naive approach is actually meaningful. Consider the free group F_r with r generators $g_1, g_2, \ldots, g_r$. Each element of F_r may be identified with a word $w_{i_1}, w_{i_2}, \ldots, w_{i_n}$, where each w_i is one of the g_j or their inverses. This representation becomes unique if we require that successive w_i not cancel each other out. To compactify F_r, we allow words of infinite length, subject, of course, to the same restriction regarding successive letters. Thus, if we set $\Omega_r = \{w = w_{i_1}, w_{i_2}, \ldots, w_{i_n}, \ldots : w_{i_n}, w_{i_{n+1}} \neq 1\}$, then in the topology of pointwise convergence, $F_r \cup \Omega_r$ forms a compact space, and Ω_r may be thought of as the boundary of F_r.

There is a way of checking this definition. Since F_2 and all its subgroups of finite index are full subgroups of the same Lie group – $SL(2, \boldsymbol{R})$ – we would expect them all to have the same boundary. What, in fact, can be shown, is that if Γ is a subgroup of finite index in F_r, then Γ is dense 'at the boundary', i.e., that $\bar{\Gamma}$ is Γ together with the entire boundary Ω_r. Suppose, for example, that $f_n \to \omega \in \Omega_r$ with $f_n \in F_r$. The length of f_n must become infinite, so that if f is a fixed element of F_r, $f_n f$ still converges to ω. Now for some f, infinitely many $f_n f \in \Gamma$, so that ω is still a limit point of Γ.

While this approach works in a few other cases as well – that is, it produces a space that at least seems a likely candidate for the boundary – it does not seem to work for the general discrete group, given, say, by finitely many generators and relations. (Incidentally, Dynkin in [8] seems to give the impression that this construction always produces the Martin boundary of a random walk on a finitely generated group. There are however some obscurities in his discussion.)

To handle the general case we find it necessary to introduce some additional structure into the picture. The objects we find we must treat are not groups *per se*, but groups together with *group invariant random walks* defined on them. (In fact we need only adjoin to the group the probability measure that defines the random walk, as we shall see later. However it will serve our purposes better for the present to think at once of the more picturesque random walk.) To be given a *random walk* means to know for each point of a space – called the *state space* – a set of probabilities for transition to another point of the space. We have in mind a particle wandering among the points of the space; at each instant its next position depends randomly – though with fixed probabilistic rules – on its present location. If the state space is a group, we say the random walk is *group invariant* if the probability of transition from g_1 to g_2 is the same as that from gg_1 to gg_2. The classical 'simple' random walk, say for dimension 2, comes about by assigning to each point (x, y) of the 2-dimensional lattice $\mathbf{Z} \times \mathbf{Z}$, the set of probabilities $\{\frac{1}{4}, \frac{1}{4}, \frac{1}{4}, \frac{1}{4}\}$ for transitions respectively to the points $(x+1, y)$, $(x-1, y)$, $(x, y+1)$, $(x, y-1)$. This is clearly group invariant. The simple m-dimensional random walk is similarly defined.

What sort of questions does one ask about random walks? We answer this by recalling a striking result from the classical theory ([9]). If U_n denotes the position of the random walk at time n (the U_n are, of course, random variables) we may ask for the probability that the U_n sometime visit a specified point of the state space. The result in question states that the simple random walk on $\mathbf{Z}$ and on $\mathbf{Z}^2$ visits every point of the state space infinitely often with probability 1, whereas for $m > 2$, with probability 1, points are visited only finitely often, and some points are not visited at all. The former case is referred to as *recurrence*, the latter as *transience*. Generally speaking, the type of question that tends to have an uncomplicated answer is one involving the 'ultimate' behavior of the random walk, that is, questions regarding asymptotic behavior of U_n for n large. Thus transience means that $U_n \to \infty$ as $n \to \infty$, and recurrence means that it does not. We emphasize that all this refers to probability 1 behavior. For the simple random walk in any number of dimensions, it is, of course, *possible* for anything to happen; it is just unlikely.

The notion of a boundary in this context enters if we now ask for morc refined behavior in the transient case. Namely, in that case, since $U_n \to \infty$, we inquire whether we might not separate out different directions along which our particle flees to ∞. The boundary is then a space adjoined to the state space of the random walk, consisting of all the possible limits of U_n as $n \to \infty$. It must be so constructed that we may assert that U_n converges with probability 1. The background for this notion comes from the theory of Brownian motion. This is a continuous version of the random walks we are discussing. Here $U(t)$ is a family of random variables defined for continuous t, taking their values in a Euclidean space, say $\boldsymbol{R}^2$ for concreteness. Each path

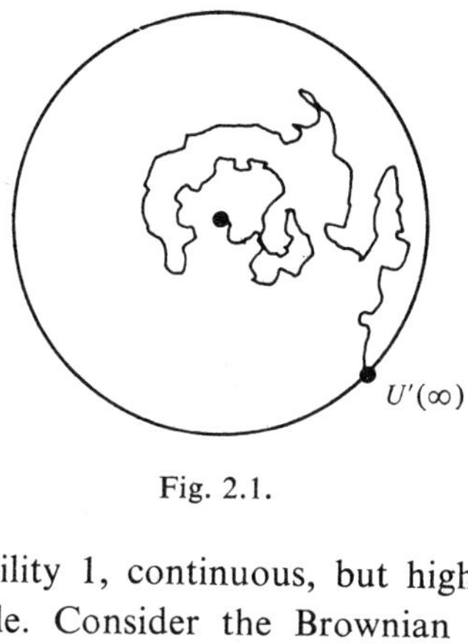

Fig. 2.1.

$U(t)$ is, with probability 1, continuous, but highly irregular, being nowhere differentiable. Consider the Brownian motion inside the unit disc. Each path ultimately hits the bounding circle of the disc. (As always, with probability 1.) Now it is possible to slow down the motion as it approaches the boundary in such a way, that if we call this modified Brownian motion $U'(t)$, then the paths of the two processes are the same, but $U'(\infty)$ is the position of $U(t)$ for the first t at which the path hits the boundary. For this diffusion, the state space is the interior of the unit disc. With this definition of the state space, the modified Brownian motion is a transient diffusion, inasmuch as the particle ultimately wanders out of every compact set of the state space. On the other hand it is also clear that that we can describe the behavior at ∞ more precisely by saying that each path tends toward a definite point of the unit circle. Of course, we could compactify the disc by merely adding a point at infinity rather than an entire circle. On the other hand, it is clearly to our advantage to obtain as precise information as possible regarding the ultimate behavior of a random walk.

Actually, these refinements do not play a role for the classical random walks. It turns out that for all $\boldsymbol{Z}^m$, irrespective of the dimension, the boundary cannot be chosen any larger than a point. To obtain interesting behavior at infinity for a group invariant random walk, we must choose a non-abelian group. For example, let $G = SL(2, \boldsymbol{R})$, and let U_n be a group invariant random walk on G subject only to the condition that U_n is not confined to a subgroup of G. (It will develop that this requirement is far too stringent.) It is possible to show – we will see this in Section 4 – that both columns of the matrix U_n tend, with probability 1, toward a fixed direction. More precisely, if we write $U_n = \left(\begin{smallmatrix} a_n & b_n \\ c_n & d_n \end{smallmatrix}\right)$, then $a_n/c_n \to t$, and $b_n/d_n \to t$, where t is a point on the projective line, i.e., t may be ∞. This shows that it is possible to distinguish different types of behavior at infinity of U_n according to the value of t occurring in the limit. Indeed, the boundary of $SL(2, \boldsymbol{R})$ with respect to any of these random walks is (subject to certain smoothness conditions) exactly $\boldsymbol{P}^1$, the one-dimensional projective space.

The boundary is then to be attached to a pair $(G, \{U_n\})$ consisting of a group and a group invariant random walk. The boundary consists of a topological space B adjoined to G to form a third topological space $G \cup B$ with the property that, with probability 1, U_n converges to a point of B. Due to the group invariance, it will suffice for studying the boundary behavior of U_n to assume that the random walk originates at the identity. For this version of the random walk we consider the distribution in B of $\lim U_n$. This determines a probability measure ν on B. Because of the importance of the latter we shall speak of B merely as being the boundary *space*, whereas the boundary will consist of the pair (B, ν). To be more specific, this pair will be termed the *Poisson boundary* (due to its role in the Poisson representation theorem for harmonic functions).

Our plan is now the following. We wish to show that a discrete group determines its envelope to some extent. In particular we claim that $SL(2, \boldsymbol{R})$ and $SL(r, \boldsymbol{R})$, $r > 2$, do not envelope the same discrete group. This will involve our looking at random walks on the three groups, $SL(2, \boldsymbol{R})$, $SL(r, \boldsymbol{R})$ and the discrete group Γ, and comparing the respective Poisson boundaries. Now for certain random walks – we will call them Brownian motions – on a semi-simple Lie group, and particularly on $SL(r, \boldsymbol{R})$, the boundary can be determined quite painlessly. For the unimodular group denote this boundary by (B_r, m_B). Our problem would be solved if we could, under the assumption that Γ is simultaneously a full subgroup of $SL(2, \boldsymbol{R})$ and a full subgroup of, say, $SL(3, \boldsymbol{R})$, produce a random walk on Γ whose boundary is at onec the same as that for the Brownian motion in $SL(2, \boldsymbol{R})$, and also the same as that for Brownian motion on $SL(3, \boldsymbol{R})$. For, the pairs (B_2, m_b) and (B_3, m_B) are very different. We shall not be able to do this however. What we shall show is that a random walk exists on Γ satisfying one of these conditions, i.e., whose boundary agrees with that of Brownian motion on one of the spaces. Inasmuch as this random walk on Γ can be viewed as taking place inside the other group as well, we will be able to conclude that while the two boundaries need not be identical, a sufficiently strong relation connects them so that we can apply a *reductio ad absurdum*. We turn now to the details.

3. Harmonic Functions and the Poisson Boundary

3.1. Probabilistic preliminaries

We give a rapid outline of the ıudiments of probability theory. A *probability space* is a triplet $(\Omega, \mathscr{B}, P)$ where Ω is a space, $\mathscr{B}$ a family of subsets of Ω including Ω and $\emptyset$ and also closed with respect to countable unions, intersections and complements, and P is a *probability measure* defined for subsets in $\mathscr{B}$. This means that for $B \in \mathscr{B}$, $P(B) \geqq 0$, $P(\Omega) = 1$, and $P(\cup B_i) = \sum P(B_i)$ whenever the B_i are disjoint sets in Ω. Generally one chooses $\mathscr{B}$ sufficiently large at the outset of the discussion so that every set that enters the discussion is *measurable*, i.e., belongs to $\mathscr{B}$. As a result one may safely ignore the reference to measurability and to the family $\mathscr{B}$. In particular, when we deal with a topological space, all the sets that appear are borel sets, and when speaking of a measure on a topological space, we simply assume from the outset that it is defined for all borel sets.

A map from Ω to a topological space X is said to be *measurable* if the inverse image of every borel set is a measurable set in Ω. Such a map is also called an *X-valued random variable.* If $X = \boldsymbol{R}$ or $\boldsymbol{C}$ we may speak of the integrability of an X-valued random variable. This integral which should be written $\int_\Omega x(\omega)\mathrm{d}P(\omega)$, is often denoted by $E(x)$, and called the *expectation* of the random variable x. A family $\{x_i\}$ of random variables is said to be independent if the probability of an event involving a number of them can be computed in terms of probabilities of events involving the variables individually according to the following formula:

$$P(x_i \in A, x_j \in B, \ldots, x_l \in D) = P(x_i \in A)P(x_j \in B) \ldots P(x_l \in D).$$

If $x_1, \ldots, x_n$ are integrable and independent, then $x_1 \ldots x_n$ is integrable, and $E(x_1, \ldots, x_n) = \prod_i E(x_i)$. More generally, suppose the x_i are independent and that the function $f(x_1, \ldots, x_n)$ is integrable. Then we can obtain its expectation by integrating in stages, treating all but one of the variables as constant at each stage. That is, if $f_1(c_2, \ldots, c_n) = E(f(x_1, c_2, \ldots, c_n))$, then $E(f(x_1, x_2, \ldots, x_n)) = E(f_1(x_2, \ldots, x_n))$, and repeating this n times, we obtain the desired expectation as an iteration of n integrations.

The passage from the random variable $f(x_1, \ldots, x_n)$ to the random variable $f_1(x_2, \ldots, x_n)$ is an example of what, under more general circumstances, is referred to as *conditional expectation*, and is written

$$f_1(x_2, \ldots, x_n) = E(f(x_1, \ldots, x_n)|x_2, \ldots, x_n).$$

The foregoing equality is then a special case of the general formula

$$E(E(z|x', x'', \ldots)) = E(z).$$

A random variable z might also be a function of infinitely many variables x_i. We needn't concern ourselves with what this means exactly; certainly if $z = \lim f_n(x_1, \ldots, x_n)$ in an appropriate sense, then this will be the case. An important result which we shall have occasion to use is

Proposition 3.1. If z is a function of $x_1, x_2, \ldots, x_n, \ldots$, and is integrable, then with probability 1, $E(z|x_1, x_2, \ldots, x_n) \to z$ as $n \to \infty$. (cf. [5]).

Taking the conditional expectation with respect to more and more variables means integrating with respect to fewer and fewer variables, so the proposition is at least plausible.

By the *distribution* of an X-valued random variable x is meant the probability measure on X defined by $\mu(A) = P(x^{-1}(A))$. Quite generally, if φ maps a space M into a space N and μ is a probability measure on M, then φ carries the measure μ into a measure ν on N in accordance with the definition $\nu(A) = \mu(\varphi^{-1}(A))$, for sets A in N. If t is a point in a space M, we denote by δ_t the probability measure concentrated at the point t: $\delta_t(A) = 1$ or 0 according as $t \in A$ or not. We see that the measure on N induced by the measure δ_t on M is just the measure $\delta_{\varphi(t)}$. This fact justifies the notation $\nu = \varphi(\mu)$ for the general case. If we think of a probability measure as a mass distribution among points of a space, then $\varphi(\mu)$ is the distribution that results on N if each point of M is moved into its image under φ. We shall often use the formula

$$\int f(s)\mathrm{d}\varphi\mu(s) = \int f(\varphi(t))\mathrm{d}\mu(t) \tag{3.1.1}$$

which is proven by showing it is valid for characteristic functions of sets, in which case it is just the definition of $\varphi\mu$.

If μ_1 and μ_2 are two measures on a space M, we say that μ, is *absolutely continuous* with respect to μ_2 if every set to which μ_2 assigns measure 0 also has μ_1-measure 0. We can then form the Radon-Nikodym derivative $\mathrm{d}\mu_1/\mathrm{d}\mu_2$ which is an integrable function with respect to μ_2, and satisfies

$$\int f(t)\frac{\mathrm{d}\mu_1}{\mathrm{d}\mu_2}(t)\mathrm{d}\mu_2(t) = \int f(t)\mathrm{d}\mu_1(t). \tag{3.1.2}$$

Now let φ be a one-one map from M to N. Then $\varphi\mu_1$ is absolutely continuous with respect to $\varphi\mu_2$. For every bounded measurable f on N we have

$$\int_N f(s)\frac{\mathrm{d}\varphi\mu_1}{\mathrm{d}\varphi\mu_2}(s)\mathrm{d}\varphi\mu_2(s) = \int_N f(s)\mathrm{d}\varphi\mu_1(s) = \int_M f(\varphi(t))\mathrm{d}\mu_1(t)$$

$$= \int_M f(\varphi(t))\frac{\mathrm{d}\mu_1}{\mathrm{d}\mu_2}(t)\mathrm{d}\mu_2(t) = \int_N f(s)\frac{\mathrm{d}\mu_1}{\mathrm{d}\mu_2}(\varphi^{-1}(s))\mathrm{d}\varphi\mu_2(s)$$

which proves that

$$\frac{d\varphi\mu_1}{d\varphi\mu_2}(s) = \frac{d\mu_1}{d\mu_2}(\varphi^{-1}(s)) \tag{3.1.3}$$

If M is a topological space, then the space of all probability measures on M itself forms a topological space in a natural way. Namely, a neighborhood of a measure μ_0 in this space is obtained by choosing a finite set of bounded continuous functions f_i on M, and a number $\varepsilon > 0$, and specifying the neighborhood as the set of all probability measures on M for which

$$\left| \int f_i d\mu - \int f_i d\mu_0 \right| < \varepsilon$$

This means that $\mu_n \to \mu$ if for each bounded continuous f on M we have $\int f(t)d\mu_n(t) \to \int f(t)d\mu(t)$. When in the sequel we speak of convergence of measures, it is always with reference to this topology. An important fact to bear in mind is that if the space M is compact, so too is the space of probability measures on M. Call this space of measures $\mathscr{M}(M)$. We observe that M can be identified with the subset of $\mathscr{M}(M)$ consisting of point measures. If M is a compact Hausdorff space, this identification is a homeomorphism.

Let $\{(\Omega_n, \mathscr{B}_n, P_n)\}$ denote a sequence of probability spaces. On the product space $\Omega = \prod_1^\infty \Omega_n$ there is defined a field $\mathscr{B}$ of sets by requiring that all the sets of the form $A_1 \times A_2 \times \ldots \times A_n \times \Omega_{n+2} \times \ldots$ are in $\mathscr{B}$. On this field we obtain a probability measure by setting $P(A_1 \times A_2 \times \ldots \times A_n \times \Omega_{n+1} \times \ldots) = \prod_{i=1}^{u} P(A_i)$. We call $(\Omega, \mathscr{B}, P)$ the *product* of the $(\Omega_n, \mathscr{B}_n, P_n)$. If x_i denotes the ith coordinate map: $\Omega \to \Omega_i$, then the x_i clearly form an independent family of random variables. We can also check that the distribution of x_i is P_i.

A particular case of this set up will play a special role in our discussion. Suppose G is a locally compact group and that μ is a probability measure on G. (Only two cases will really come under consideration: G a group of matrices and G a discrete group. In the latter case μ corresponds to a sequence of non-negative numbers $\mu(g)$ with $\sum \mu(g) = 1$.) We form the infinite product

$$(\Omega, \mathscr{B}, P) = (G, \mathscr{B}_G, \mu) \times (G, \mathscr{B}_G, \mu) \times \ldots \times (G, \mathscr{B}_G, \mu) \times \ldots$$

where $\mathscr{B}_G$ denotes the field of borel sets on G. Just about all the probability theory we will be doing takes place on this space $(\Omega, \mathscr{B}, P)$.

We again let $\{x_i\}$ denote the sequence of coordinate functions. We shall speak of this sequence as a *stationary sequence of independent r.v.'s with distribution* μ.

Now form the product random variables $u_n = x_1 x_2 \ldots x_n$. The sequence $\{u_n\}$ forms what is called a *Markov process*. We content ourselves with an informal definition of the latter. A sequence of random variables $\{y_n\}$ forms a *Markov process* if the probabilistic structure of the process is entirely determined by specifying the probabilities of the transitions from y_n to y_{n+1}, the probability of a sequence of steps $y_1 \to y_2 \to y_3 \to \ldots y_n$ being the product of the probabilities of the individual transitions $y_i \to y_{i+1}$. (When the state space is not discrete these probabilities generally all vanish, and a meaningful definition must be made in terms of integrals.) Loosely speaking, this means that having arrived at a position in the state space, the particle that undergoes a Markov process forgets its entire past and proceeds according to a probabilistic rule that depends only on this position. In our case we see that this is indeed so. For the transition from u_n to u_{n+1} proceeds by right multiplication with x_{n+1}. Since x_{n+1} is independent of the x_i for $i \leqq n$, and therefore of the u_i, $i \leqq n$, it follows that once we know u_n, the u_i for $i < n$ have nothing to contribute to the determination of u_{n+1}. A Markov process on a group for which the transitions consist of group multiplication is generally called a *random walk*. As far as we are concerned all this merely provides us with terminology. We know as much as we need to know regarding the sequence $\{u_n\}$ by specifying that they represent the partial products in G of the variables in the sequence $\{x_n\}$. We remark that the entire setup is determined once we are given G and the probability measure μ.

3.2. G-SPACES, EQUIVARIANT MAPS, AND CONVOLUTIONS

A G-space is a topological space on which the group G acts continuously. More precisely, a space T is a G-space if there is defined a map $G \times T \to T$ such that, denoting the image of (g, t) by gt, we have $et = t$ for the identity e of G, and $(g_1 g_2)t = g_1(g_2 t)$.

Let x be a G-valued r.v. and let y be a T-valued r.v., where T is a G-space. The variable xy is then also a T-valued r.v. Assume x and y are independent and that they have distributions μ and ν respectively.

As a result of the independence, the distribution of xy is seen to depend only on μ and ν. This distribution, a probability measure on T, will be denoted $\mu * \nu$. Recalling our remarks from Section 3.1, we see that

$$\int f(t) \mathrm{d}\mu * \nu(t) = \int\int f(gt) \mathrm{d}\nu(t) \mathrm{d}\mu(g). \tag{3.2.1}$$

When μ is a point measure, $\mu = \delta_g$, we shall abbreviate the measure $\mu * \nu$ to $g\nu$. It agrees with the image of ν under g, $g(\nu)$, defined in the foregoing section.

G is, of course, itself a G-space, $(g_1, g_2) \to g_1 g_2$, so the *convolution* $\mu * \nu$ is defined when both μ and ν are probability measures on G.

Let T and S both be G-spaces for a group G. A map $f: T \to S$ is said to be *equivariant* if $f(gt) = gf(t)$ for all $t \in T$ and $g \in G$. Let us illustrate this notion. Let G be a group of $r \times r$ matrices, $T = \boldsymbol{R}^r - \{0\}$, and $S = \boldsymbol{P}^{r-1}$, the $r-1$-dimensional projective space. Both T and S are G-spaces since G acts both on points of R^r as well as on the lines through the origin of R^r. Now define $f: T \to S$ by sending each point of R^r other than the origin into the line that passes through the point and the origin. It is easy to see that f is an equivariant map.

Now suppose that μ is a probability measure on G, ν a probability measure on the G-space T, and $f(\nu)$ its image on the G-space S by the equivariant map f. Let μ correspond to the variable x and ν to the variable y. $\mu * \nu$ is the distribution of the variable xy, and we have $f(xy) = xf(y)$. It follows that for equivariant maps

$$f(\mu * \nu) = \mu * f(\nu). \tag{3.2.2.}$$

3.3. μ-Processes

We turn now to more special notions that will be useful in our discussion of the Poisson boundary. Let $\{x_n\}$ denote as before a stationary G-valued sequence of independent r.v.'s with distribution μ and form the products u_n. We recall that the Poisson boundary was to be a space B which could be adjoined to the group G to form a space $G \cup B$ with the property that a certain 'group invariant random walk' converges, with probability 1, to points in B. The random walks we had in mind are those of the form gu_n, g an arbitrary element of G. All group invariant random walks that originate at a specified point g of the group are, in fact, of this form. Thus we wish to construct a space B, and attach B to G so that, with probability 1, gu_n converges to some point of B. Assume now that B is chosen to be a G-space, and that, moreover, B is attached to G in such a way that the union $G \cup B$ remains a G-space. Then if u_n converges with probability 1, so too will gu_n, and the limits of the latter are determined by those of the former. In particular, if $u_n = x_1 x_2 \ldots x_n$ converges, say to the B-valued random variable z_1, then each of the sequences $x_k x_{k+1} \ldots x_{k+n}$ converges. If we denote the limit of the latter by z_k, we find that we have defined in terms of $\{x_n\}$ a new sequence of random variables with value sin the G-space B, and satisfying

(a) z_k is a function of $x_k, x_{k+1}, \ldots$

(b) all the z_k have the same distribution,

(c) x_k is independent of $z_{k+1}, z_{k+2}, \ldots$

(d) $z_k = x_k z_{k+1}$.

(c) is a consequence of (a) and the fact that the x_n themselves formed an independent sequence.

Definition 3.1. A sequence $\{z_n\}$ of random variables with values in a G-space which is defined together with the sequence $\{x_n\}$ and satisfying (a)-(d) is called a μ-process.

Let ν denote the distribution of z_1. By (c) and (d) we have $\mu * \nu = \nu$.

Definition 3.2. A probability measure on a G-space is called *stationary for* μ if $\mu * \nu = \nu$.

Notice that by (3.2.2), an equivariant image of a stationary measure is stationary. For, if ν is stationary, then

$$\mu * f(\nu) = f(\mu * \nu) = f(\nu).$$

Returning to the space B above, we see that in order for a G-space to qualify as the Poisson space relative to the random walk $\{x_n\}$, it must at least possess a stationary measure for μ. Let us notice now that if we know the measure ν, then we can essentially determine the topological structure of the union $G \cup B$. For we have

Proposition 3.2. If $\{z_n\}$ is a μ-process and ν is the common distribution of the z_n, then with probability 1,

$$x_1 x_2 \ldots x_n \nu \to \delta_{z_1}. \tag{3.3.1}$$

Note that the product in (3.3.1) is meaningful, being the product of an element of G and a measure in a G-space.

Proof. We must show that for a bounded continuous function f on the G-space in question – call it B –

$$\int_B f(\xi)\mathrm{d}x_1 x_2 \dots x_n \nu(\xi) = \int_B f(x_1 x_2 \dots x_n \xi)\mathrm{d}\nu(\xi) \to f(z_1).$$

Let us evaluate $E(f(z_1)|x_1, x_2, \dots, x_n)$. Writing z_1 as $x_1 x_2 \dots x_n z_{n+1}$, and observing that the $n+1$ factors here are independent by (c), we find that taking the conditional expectation in question entails an integration with respect to the variable z_{n+1} which has the distribution ν. We then find

$$E(f(z_1)|x_1, x_2, \dots, x_n) = E(f(x_1 x_2 \dots x_n z_{n+1})|x_1, x_2, \dots, x_n)$$
$$= \int f(x_1 x_2 \dots x_n \xi)\mathrm{d}\nu(\xi).$$

But by Proposition 3.1, the conditional expectation converges to $f(z_1)$, and this proves our proposition.

This shows us how, given the space B and the measure ν, we may retrieve the limit relation $x_1 x_2 \dots x_n \to z_1$. Namely, if we identify the points of B with the point measures they support, and if we consider a sequence g_n in G convergent to a measure of $\mathcal{M}(B)$ whenever the sequence of measures $g_n \nu$ converges in $\mathcal{M}(B)$ to the measure in question, then Proposition 3.2. says exactly that $x_1 x_2 \dots x_n$ converges to z_1. An alternative, more precise definition of the topology in $G \cup B$ is this:

Definition 3.3. If B is a G-space and ν a probability measure on B, the *measure topology of $G \cup B$ with respect to ν* is defined as the weakest topology for which the imbeddings $G \to G \cup B$, $B \to G \cup B$ are homeomorphisms into, and for which the map of $G \cup B$ into $\mathcal{M}(B)$ defined by $g \to g\nu$, $\xi \to \delta_\xi$ is continuous.

What we have found is this. A space B of the kind we are looking for leads to a μ-process. But conversely, if there is a μ-process with values in the G-space B, then B is a space of the type that we are looking for. For then, placing the measure topology on $G \cup B$ with respect to the distribution of the μ-process variable, we find, by Proposition 3.2, that our random walk does indeed converge to a 'boundary' point.

The next question to be asked is this. Suppose B is a G-space, under what conditions does there exist a B-valued μ-process? The answer is given in the next proposition.

Proposition 3.3. Let B be a G-space and ν a stationary probability measure on B, i.e., $\mu * \nu = \nu$. There exists a B-valued μ-process whose variables have distribution ν if and only if with probability 1, the measures $x_1 x_2 \ldots x_n \nu$ converge to point measures on B.

Proof. The necessity has already been shown. For the sufficiency, set $\delta_{z_k} = \lim x_k x_{k+1} \ldots x_{n+k}\nu$. Conditions (a), (c), and (d) are immediate for the process so defined. Condition (b) also is evident, for the sequence $x_k, x_{k+1}, \ldots$ is probabilistically isomorphic to the sequence $x_1, x_2, \ldots$. We have only to check that the distribution of z_1 is ν. But we have

$$E(f(z_1)) = \lim_{n\to\infty} E\left(\int f(x_1 x_2 \ldots x_n \xi) d\nu(\xi)\right)$$

$$= \lim_{n\to\infty} \int f(\xi) d\overbrace{\mu * \mu * \ldots * \mu}^{n} * \nu(\xi) = \int f(\xi) d\nu(\xi)$$

by the stationarity of ν. This proves the proposition.

Suppose now that B and B' are two G-spaces, and that B' is an equivariant image of B, $\rho: B \to B'$. Suppose B supports a μ-process $\{z_n\}$. It is clear that the variables $z_n = \rho(z_n)$ define a B'-valued μ-process. Introduce the following terminology:

Definition 3.4. A G-space B together with a stationary measure ν is said to form a *boundary* of (G, μ) if ν is the distribution of a μ-process on B. We say a boundary (B', ν') is an equivariant image of the boundary (B, ν) if there is an equivariant map ρ that takes $B \to B'$ and $\rho\nu = \nu'$.

Recall from Section 2 that the Poisson boundary was to be chosen as large as possible in order to give as much information as possible regarding the limit behavior of u_n. In particular, we now see that the Poisson boundary should be a boundary in the sense of Definition 3.4 which is not at the same time a 'proper' equivariant image of some other boundary. We will discover in Section 3.5 that among the boundaries of (G, μ) there always exists a maximal one in the sense that all other boundaries are equivariant images of this one. Clearly that should be our choice for the Poisson boundary.

3.4. μ-HARMONIC FUNCTIONS

One more ingredient is necessary before we can make precise our definition of the Poisson boundary.

Definition 3.5. Let $h(g)$ be a bounded measurable function on the group G. h is said to be μ-harmonic if it satisfies the mean value property

$$h(g) = \int h(gg')\mathrm{d}\mu(g'). \tag{3.4.1}$$

Letting $\{x_n\}$ denote as before a stationary sequence of independent r.v.'s with distribution μ, we find that (3.4.1) can be rewritten as

$$h(g) = E(h(gx_n)). \tag{3.4.2}$$

The following is a probabilistic analogue of the Fatou theorem regarding boundary behavior of harmonic functions. This proposition suggests that μ-harmonic functions play an important role in defining an appropriate boundary for the random walk u_n.

Proposition 3.4. For a fixed $g \in G$, let $w_n = h(gx_1 x_2 \ldots x_n)$ with $w_0 = h(g)$. With probability 1, w_n converges as $n \to \infty$, and setting $w_\infty = \lim w_n$, we have $E(w_\infty) = h(g)$.

Proof. This is a consequence of the rather deep 'martingale convergence theorem' ([5]). We shall show how it may be derived from Proposition 3.1. Let $y_n = w_n - w_{n-1}$ with $y_0 = w_0$. (3.4.2) implies that $E(h(gx_1 x_2 \ldots x_n)|x_1, x_2, \ldots, x_{n-1})$ is $h(gx_1 x_2 \ldots x_{n-1})$. Thus $E(y_n|x_1, x_2, \ldots, x_{n-1}) = 0$. Now let us evaluate $E(y_n \bar{y}_{n-i})$ for $i > 0$. We first perform the integration with respect to x_n, leaving the other variables fixed. This leaves $\bar{y}_{n-i}$ fixed, and the result is proportional to $E(y_n|x_1, x_2, \ldots, x_{n-1}) = 0$. Hence in the Hilbert space $L^2(\Omega, P)$, the variables y_n form orthogonal elements. Moreover

$$||y_0 + y_1 + \ldots + y_n||_2 = ||w_n||_2 \leqq ||h||_\infty.$$

It follows that $w_n = y_0 + y_1 + \ldots y_n$ converges in $L^2(\Omega, P)$ to some variable w_∞. Now it is easily seen that if $w_n \to w_\infty$ in $L^2(\Omega, P)$ then $E(w_n|x_1, x_2, \ldots, x_k)$ converges to $E(w_\infty|x_1, x_2, \ldots x_k)$. On the other hand, since $E(y_j|x_1, x_2, \ldots, x_k) = 0$ for $j > k$, we have $E(w_n|x_1, x_2, \ldots x_k) = w_k$ for $n \geqq k$. It follows that $E(w_\infty|x_1, x_2, \ldots, x_k) = w_k$.

We can now apply Proposition 3.1 to conclude that w_k converges to w_∞ with probability 1. Finally $E(w_\infty) = \lim E(w_k) = h(g)$ by repeated application of (3.4.2).

An immediate connection between μ-harmonic functions and boundaries is given by the following.

Lemma 3.1. Let ν be a stationary measure for μ on a G-space B. Then the function

$$h(g) = \int_B \varphi(g\xi)\mathrm{d}\nu(\xi)$$

is a μ-harmonic function on G, for any bounded measurable φ on B.

Proof. For

$$\begin{aligned} h(g) &= \int_B \varphi(g\xi)\mathrm{d}\nu(\xi) = \int_B \varphi(g\xi)\mathrm{d}\mu * \nu(\xi) \\ &= \int_G \int_B \varphi(gg'\eta)\mathrm{d}\mu(g')\mathrm{d}\nu(\eta) = \int h(gg')\mathrm{d}\mu(g'). \end{aligned}$$

Corollary. If B is a compact Hausdorff space with a countable base, and if ν is a stationary measure for μ on B, then the sequence of measures $x_1 x_2 \ldots x_n \nu$ is convergent in $\mathcal{M}(B)$, with probability 1.

Proof. If φ is bounded and continuous in B, then

$$\int \varphi(\xi)\mathrm{d}x_1 x_2 \ldots x_n \nu(\xi) = \int \varphi(x_1 x_2 \ldots x_n \xi)\mathrm{d}\nu(\xi) = h(x_1 x_2 \ldots x_n)$$

and this converges with probability 1 according to Proposition 3.4. Now if B has a countable base, then there exists a countable family of continuous functions dense in the space of all continuous functions on B. If we have convergence of a sequence of measures in $\mathcal{M}(B)$ relative to a dense subset of continuous functions, then the sequence must actually converge. Since we have to contend only with a countable family of functions, the union of all the exceptional sets that must be taken into consideration still forms a set of probability 0.

3.5. The Poisson boundary

The principal tool in our subsequent discussion is the following existence theorem. ([11, p. 372] for the Lie group case.)

Theorem 3.1. If G is a locally compact group satisfying the second axiom of countability (G has a countable basis for its open sets), and μ is a probability measure on G, then there exists a boundary (B, ν) for (G, μ) with the following properties:

i) Every boundary of (G, μ) is an equivariant image of (B, ν).

ii) If $h(g)$ is any μ-harmonic function on G, there exists a bounded measurable function $\hat{h}(\xi)$ on B such that $h(g) = \int_B \hat{h}(g\xi) d\nu(\xi)$. The function $\hat{h}$ is unique in the following sense. Call a subset A in B a *nullset* if $\nu(gA) = 0$ for each $g \in G$. A *nullfunction* is one that vanishes outside a nullset. Then the *boundary function* $\hat{h}(\xi)$ is uniquely determined by the harmonic function $h(g)$ modulo nullfunctions.

The boundary referred to in this theorem is called the *Poisson* boundary and is denoted $(B, \nu) = P(G, \mu)$.

The proof is tedious and technical. As a result we invite – even urge – the impatient reader to skip the remainder of this section.

Proof. To motivate the method of proof, let us assume for a moment that the space B has already been constructed. From previous discussions we know that there is determined a topology on $G \cup B$ such that the random walk $x_1 x_2 \ldots x_n$ converges with probability 1 on $G \cup B$. So if $f(g)$ is a function on G which extends as a continuous function to all of $G \cup B$, then $f(gx_1 x_2 \ldots x_n)$ converges with probability 1. It is therefore quite natural in attempting to produce the boundary space B, to start with the class of all functions on G that have this property. So let $\mathscr{A}$ denote this class of functions: $f \in \mathscr{A}$ if for each $g \in G$, $f(gx_1 \ldots x_n)$ converges with probability 1. According to Proposition 3.4, the class $\mathscr{H}$ of μ-harmonic functions on G is contained in $\mathscr{A}$. $\mathscr{A}$ is a Banach algebra with respect to the ess. sup. norm, whereas $\mathscr{H}$ has no obvious multiplication defined on it.

We define two other spaces of functions. Let $\mathscr{J}$ denote the closed ideal in $\mathscr{A}$ of functions f for which $f(gx_1 \ldots x_n)$ converges to 0 with probability 1. Also let $\mathscr{L}$ denote the space of limit functions, $z_g = \lim f(gx_1 \ldots x_n), f \in \mathscr{A}$. That is to say, each element of $\mathscr{L}$ is a random function on G, given by the limiting expression above. $\mathscr{L}$ is again an algebra and quite obviously we have

$$\mathscr{L} \cong \mathscr{A}/\mathscr{J}. \tag{3.5.1}$$

However, there is also an isomorphism between $\mathscr{H}$ and $\mathscr{A}/\mathscr{J}$. In fact we have

Lemma 3.2. $\mathscr{A} = \mathscr{H} + \mathscr{J}$.

Proof. Let $f \in \mathscr{A}$. Set $z_g = \lim f(gx_1 \ldots x_n)$, and let $h(g) = E(z_g)$. We claim h is μ-harmonic. For, let x_0 be a random variable, independent of all the x_i, $i \geqq 1$, and with distribution μ. Then the system $x_0, x_1, \ldots$ is isomorphic to the system $x_1, x_2, \ldots$. As a result,

$$h(g) = E(\lim f(gx_0 x_1 \ldots x_n)) = E(h(gx_0))$$

which proves our assertion. Consider next $f - h$. Again shifting variables we see that

$$\begin{aligned} h(gx_1 \ldots x_n) &= E(\lim f(gx_1 \ldots x_n x_{n+1} \ldots) | x_1, \ldots, x_n)) \\ &= E(z_g | x_1, \ldots, x_n). \end{aligned}$$

Hence $\lim h(gx_1 \ldots x_n) = z_g = \lim f(gx_1 \ldots x_n)$ and $f - h \in \mathscr{J}$. Thus each function f in $\mathscr{A}$ is a sum of a μ-harmonic function and a function in $\mathscr{J}$. Moreover $\mathscr{H} \cap \mathscr{J} = 0$ since, by Proposition 3.4, a μ-harmonic function may be retrieved from its limit function. This proves the lemma.

In other words we have

$$\mathscr{H} \cong \mathscr{A}/\mathscr{J} \cong \mathscr{L}. \tag{3.5.2}$$

Now $\mathscr{A}/\mathscr{J}$ is a commutative C^* algebra, and therefore admits a (Gelfand) representation as the space of continuous functions on a compact Hausdorff space B'. Thus we have a $1-1$ correspondence established between μ-harmonic functions $h(g)$ and continuous functions $\tilde{h}(\eta)$ on B'. Moreover, the linear functional sending a μ-harmonic function to its value at e: $h \to h(e)$, is non-negative for non-negative elements of $\mathscr{A}/\mathscr{J}$ and sends the function 1 into the value 1. It must therefore, by the Riesz representation theorem, be given by

$$h(e) = \int_{B'} \tilde{h}(\eta) d\nu'(\eta)$$

for a uniquely determined probability measure ν' on B'.

We can also define the action of G on B'. For the operations $f(g) \to f(g'g)$ send $\mathscr{A}$ into $\mathscr{A}$ and $\mathscr{J}$ into $\mathscr{J}$, and so they define a group of automorphisms of $\mathscr{C}(B')$. Such a group always corresponds to a group of homeomorphisms of B'. Thus B' is a G-space and if

$h(g)$ corresponds to $\tilde{h}(\eta)$, then $h(g'g)$ corresponds to $\tilde{h}(g'\eta)$. Incidentally, this gives

$$h(g) = \int \tilde{h}(g\eta)\mathrm{d}\nu'(\eta). \tag{3.5.3}$$

Up till this point our construction has involved canonically defined objects, and the space B' that it leads to is uniquely determined. Unfortunately the space B' is somewhat too large for our purposes. In general it will not possess a countable base and this makes it unnatural to speak of convergence of sequences of points or measures on B'. More specifically, the problem is this. To prove that a sequence of measures θ_n on B' converges with probability 1, we must show that outside of an exceptional set of probability 0, $\int\psi(\eta)\mathrm{d}\theta_n(\eta)$ converges for *each* continuous ψ on B'. If the space had a countable base, we may choose a countable dense set in $\mathscr{C}(B')$, and then it suffices to show that the exceptional set of nonconvergence of $\int\psi(\eta)\mathrm{d}\theta_n(\eta)$ for a *particular* ψ has probability 0.

We get around this as follows. The group G being separable, it follows that the spaces $L^p(B', \nu')$ are separable. We can therefore find a subalgebra of $L^\infty(B', \nu')$ which does have a countable dense set, and which is also dense in each of the L^p-spaces, $p < \infty$. Moreover, inasmuch as G has a countable dense set, the algebra may be chosen G-invariant. This subalgebra defines an identification space B of B' so that $\mathscr{C}(B)$ is isomorphic to the algebra in question. Let ν be the image of ν' under the natural map $B' \to B$. The space $L^p(B, \nu)$ is the completion of $\mathscr{C}(B)$ in the L^p norm of ν, and since this algebra corresponds to a dense subalgebra of $L^p(B', \nu')$, it follows that the L^p-spaces in question are the same. Thus while in passing from B' to B, the class of continuous functions is made smaller, the class of bounded measurable functions remains the same.

The Poisson representation (3.5.3) for the bounded μ-harmonic functions on G now becomes

$$h(g) = \int \hat{h}(g\xi)\mathrm{d}\nu(\xi), \tag{3.5.4}$$

with $\hat{h}$ bounded and measurable, though no longer necessarily continuous. On the other hand, starting with a continuous boundary function $\hat{h}(\xi)$ and forming the harmonic function $h(g) = \int\hat{h}(g\xi)\mathrm{d}\nu(\xi)$

we find that ν and $\mu * \nu$ define the same linear functional on $\mathscr{C}(B)$. Hence ν is a stationary measure for μ.

We wish to show that (B, ν) is a boundary of (G, μ). By Proposition 3.3, this will be the case if $x_1 \ldots x_n \nu$ converges on B to a point measure, with probability 1. By the Corollary to Lemma 3.1, we know that the sequence of measures in question does converge to something. Say $x_1 \ldots x_n \nu \to \theta$. To prove that θ is, with probability 1, a point measure, fix a function $\varphi \in \mathscr{C}(B)$, and consider the corresponding μ-harmonic function $h(g)$. Applying (3.5.4), we obtain

$$\int \varphi(g\xi) \mathrm{d}\theta(\xi) = \lim \int \varphi(gx_1 x_2 \ldots x_n \xi) \mathrm{d}\nu(\xi) = \lim h(gx_1 \ldots x_n) = z_g,$$

where z_g is the random function in $\mathscr{X}$ corresponding to h. Now the correspondence between random functions of $\mathscr{X}$ and boundary functions on B is an algebra isomorphism, and according to the foregoing, the relation between the two spaces can be expressed by

$$z_g = \int \varphi(g\xi) \mathrm{d}\theta(\xi).$$

This being an algebra isomorphism, we must have

$$z_g^2 = \int \varphi(g\xi)^2 \mathrm{d}\theta(\xi) \geqq \left\{ \int \varphi(g\xi) \mathrm{d}\theta(\xi) \right\}^2 = z_g^2$$

Equality can take place here only if $\varphi(\xi)$ is constant with respect to θ, and since this has to happen for all $\varphi \in \mathscr{C}(B)$, it follows that θ is a point measure.

We have shown that a boundary exists which produces all the μ-harmonic functions on G in the manner required by the theorem. The uniqueness assertion for $\hat{h}$ can be obtained by showing that the nullfunctions in the theorem are exactly those that correspond to vanishing μ-harmonic functions. We omit the details inasmuch as this assertion will not play any role in the sequel. What still remains to be shown is that (B, ν) is the *maximal* boundary, i.e., that any other boundary is an equivariant image of this one.

To this end consider a boundary (B_1, ν_1). We shall show that there is a C^*-algebra isomorphism of $\mathscr{C}(B_1)$ into $L^\infty(B)$ which is compatible with the action of G on the two spaces, and such that the linear functionals defined on these two spaces by ν_1 and ν respectively correspond under the isomorphism.

Let $\varphi \in \mathscr{C}(B_1)$ and form the μ-harmonic function $h(g) = \int \varphi(g\xi)\mathrm{d}\nu'(\xi)$. Let z_g denote the corresponding limit function in $\mathscr{Z}$. If we denote by $\{v_n\}$ the μ-process on B_1, then we have, by Proposition 3.2, $x_1 \ldots x_n\nu' \to \delta_{v_1}$, and so

$$\varphi(gv_1) = \lim \int \varphi(gx_1 \ldots x_n\xi)\mathrm{d}\nu'(\xi) = \lim h(gx_1 \ldots x_n) = z_g.$$

According to this formula, the map $\varphi \to z_g$ is an algebra isomorphism, and since $\mathscr{L}$ is isomorphic to $L^\infty(B)$ modulo nullfunctions, we find that $\mathscr{C}\ (B_1)$ is isomorphic to a subalgebra of the latter. The action of G is clearly preserved. Moreover the functionals corresponding to both the measures ν_1 and ν are given by $z_g \to E(z_e)$. This completes our proof.

It is clear from our proof that as a topological space, the Poisson boundary is not uniquely determined. What is uniquely determined is a measure-theoretic object – the algebra $L^\infty(B, \nu)$ modulo nullfunctions. Nonetheless, we do not dispense entirely with the topological structure of B, because without it, we lose the significance of the Poisson boundary as a boundary attached to the topological group.

In concluding this section a few words should be said regarding the so-called *Martin boundary*. For more details we refer the reader to [6]. The Martin boundary is a topological space related to the class of positive (not necessarily bounded) harmonic functions in much the same manner as the Poisson boundary is related to the class of bounded harmonic functions. The boundary 'values' in the case of the Martin boundary, however, are measures and not functions. It can be shown that the Martin boundary of (G, μ) always constitutes a realization of the Poisson boundary, though not vice-versa. For our purposes the Martin boundary is less useful, the reason being that it is not sufficiently flexible to enable us to obtain partial information regarding it without determining it exactly. The advantage of the Poisson boundary is that perturbing the measure μ does not appear to radically change the nature of the Poisson boundary.

4. Examples

For many groups G, the Poisson boundary $P(G, \mu)$ is trivial for every choice of the measure μ. This is so, for example, if G is abelian. To say that $P(G, \mu)$ is trivial amounts to saying that the only bounded μ-harmonic functions are the constants (which, of course, are always μ-harmonic). Now it is known ([4]) that on an abelian group, a bounded function $f(g)$ can satisfy

$$(g) = \int f(gg')\mathrm{d}\mu(g')$$

only if it is constant, and this implies the triviality of the Poisson boundary.

When the boundary is non-trivial, it is usually very difficult to determine. There are some exceptions which we discuss in this section.

4.1. The free group with two generators

In (2) we showed how to attach to the free group with r generators, F_r, a 'boundary' that compactified the group. This space may, in fact, be taken as the underlying space of the Poisson boundary for the 'simple' random walk on F_r. We consider F_2 with generators a, b. The simple random walk corresponds to the probability distribution μ on F_2 given by $\mu(a) = \mu(b) = \mu(a^{-1}) = \mu(b^{-1}) = \frac{1}{4}$. To compactify F_2 we adjoined to it the space Ω_2 of all infinite sequences $\omega = \omega_1\omega_2\omega_3 \ldots$ where each $\omega_i = a, b, a^{-1}$, or b^{-1} and no product $\omega_i\omega_{i+1}$ gives the identity. The space Ω_2 with an appropriate measure ν will then be the Poisson boundary of (F_2, μ).

Let us show that the space Ω_2 is an F_2-space, i.e., that the group F_2 acts on Ω_2 continuously. To define the action of F_2 on Ω_2 we simply proceed by continuity: $g\omega = \lim gg_n$, for a sequence $g_n \to \omega$. Writing g itself as a product, $g = \omega_1'\omega_2' \ldots \omega_l'$, we see that $g\omega$ is obtained by juxtaposing $\omega_1'\omega_2' \ldots \omega_l'$ and ω, and then collapsing if necessary. For example, if $\omega = bababa\ldots$, then $a\omega = ababab\ldots$, if $\omega = a^{-1}ba^{-1}b\ldots$, then $a\omega = ba^{-1}b\ldots$.

Take x_n to be a stationary sequence of independent F_2-valued r.v.'s with distribution μ, (in fact, x_n must then take on one of the 4 values a, b, a^{-1}, b^{-1}), then it is clear how we obtain convergence of $x_1x_2\ldots x_n$ to a point of Ω_2. We simply write out the product $x_1x_2\ldots x_nx_{n+1}\ldots$ and cross out all neighboring pairs x_ix_{i+1} that happen to cancel out; we continue this procedure until a sequence of Ω_2 has been obtained. That the product cannot keep degenerating indefinitely (outside of a set with probability 0) is guaranteed by the fact that the random walk $x_1 x_2 \ldots x_n$ is transient, i.e., it wanders out to ∞. If we thus define the *length* of an element $g \in F_2$ to be the minimal length of the product expressing g in terms of the generators and their inverses, the transience of the random walk implies that the length of $x_1 x_2 \ldots x_n$ tends to infinity. If n_k is the largest value of n for which the product has length k, then once $x_1 x_2 \ldots x_{n_k}$ is *reduced*, multiplication on the right by x_{n_k+1}, etc., will not affect the first k letters of the product. We thus obtain as limit a well-defined Ω_2-valued r.v. which we may denote unambiguously

$$z_1 = x_1 x_2 \ldots x_n \ldots.$$

Similarly we may set $z_k = x_k x_{k+1} \ldots x_{k+n} \ldots$, and $\{z_k\}$ is a sequence of Ω_2-valued r.v.'s. Clearly the z_k form a μ-process (Definition 3.1), and so if ν denotes the distribution of z_1, (Ω_2, ν) will be a boundary of (G, μ). The measure ν is not readily computed directly; however it is easy to guess what it might be. If we make this guess, and verify that the measure in question is stationary (i.e., that $\mu * \nu = \nu$), it will follow that that measure is the one sought. We see this as follows. Fix $\omega \in \Omega_2$, and form the products $x_1 x_2 \ldots x_n \omega$. It is not hard to see that with probability 1, this converges in Ω_2 to z_1. This being the case for each $\omega \in \Omega_2$, it also follows for measures, that is, for any measure ν on Ω_2, $x_1 x_2 \ldots x_n \nu$ converges to δ_{z_1}. In particular, if ν is stationary, we can apply Proposition 3.3 to conclude that there exists a μ-process with distribution ν, and then by Proposition 3.2, it follows that z_k is the μ-process in question. Hence there can exist on Ω_2 only one stationary measure for μ and it gives the distribution of the variables z_k.

Our guess for ν is that it is the measure on Ω_2 for which the coordinates have maximum independence, subject to the condition that $\omega_i \omega_{i+1}$ does not degenerate. This defines a Markov process on the space $\{a, b, a^{-1}, b^{-1}\}$: ω_1 takes on the four values with equal probabilities, ω_2 takes the three possible values it may now have with equal probabilities, and so on. This gives us

$$\nu\{\omega: \omega_1 = \omega_1^0, \omega_2 = \omega_2^0, \ldots, \omega_n = \omega_n^0\} = \frac{1}{4 \cdot 3^{n-1}}.$$

This defines ν and it is an easy matter to show that ν satisfies $\mu * \nu = \nu$. The boundary (Ω_2, ν) now has an explicit form.

In fact, (Ω_2, ν) is the Poisson boundary. We shall not prove this but refer the reader to [8] where it is shown that Ω_2 is the Martin boundary of (G, μ). As we have indicated, this implies that (Ω_2, ν) is the Poisson boundary. Alternatively the reader may test his later understanding of Section 5.2 by showing that the method to be developed there for proving that a certain boundary is the Poisson boundary, works as well in the present situation.

4.2. A LAW OF LARGE NUMBERS FOR THE FREE GROUP

One use for the boundary is in obtaining 'laws of large numbers' for random products in the group. This has been described in detail in [12] for the case that G is a Lie group. Here we shall give an application in the case of F_2. Denote by $l(g)$ the length function discussed in the last paragraph: $l(g)$ is the length of the minimal word expressing g in terms of generators and their inverses. We are interested in the asymptotic behavior of $l(x_1 x_2 \ldots x_n)$ as $n \to \infty$.

To begin with, let us note that for each g in G, the measure $g\nu$ is absolutely continuous with respect to ν. For

$$\nu = \tfrac{1}{4}(a\nu + b\nu + a^{-1}\nu + b^{-1}\nu)$$

so that our assertion is valid for g one of the generators or its inverse. But then it is also valid for any g. Actually it is easy to compute the Radon-Nikodym derivative of $g\nu$ with respect to ν. By induction on the length of g, it can be shown that if $\omega \in \Omega_2$ is the limit of g_n in F_2, then

$$\frac{dg^{-1}\nu}{d\nu}(\omega) = 3^{l(g_n) - l(gg_n)} \tag{4.2.1}$$

for sufficiently large n. Thus

$$l(gg_n) - l(g_n) \to -\log \frac{dg^{-1}\nu}{d\nu}(\omega)/\log 3$$

as $g_n \to \omega$. Consider

$$\frac{l(x_1 \ldots x_n)}{n} = \frac{1}{n}\sum_{i=1}^{n} \{l(x_i x_{i+1} \ldots x_n) - l(x_{i+1} \ldots x_n)\}$$

Now for each i

$$l(x_i x_{i+1} \dots x_n) - l(x_{i+1} \dots x_n) \to -\log \frac{dx_i^{-1}\nu}{d\nu}(z_{i+1})/\log 3.$$

From this one finds (a few steps have to be filled in here!) that

$$\frac{l(x_1 \dots x_n)}{n} \to \lim_{n\to\infty} \frac{1}{n}\sum_1^n -\log \frac{dx_i^{-1}\nu}{d\nu}(z_{i+1})/\log 3.$$

The limit on the right exists by the ergodic theorem and is given by

$$\frac{1}{\log 3}\int_{F_2}\int_{\Omega_2} -\log \frac{dg^{-1}\nu}{d\nu}(\omega)d\mu(g)d\nu(\omega).$$

Using (4.2.1) we may evaluate the integral and we obtain

$$l(x_1 x_2 \dots x_n) \sim n/2. \tag{4.2.2}$$

4.3. A RANDOM WALK ON F_∞

The method of the preceding paragraphs applies equally well to all the free groups F_r, $r < \infty$. For F_∞ the simple random walk is not defined, since there are infinitely many generators. We shall describe however a random walk on F_∞ whose Poisson boundary is 'computable'. In fact, for some measure μ' on F_∞, $P(F_\infty, \mu') = P(F_2, \mu)$, where μ is the measure of paragraph A. We are not able to exhibit the measure μ' explicitly; what we shall do is prove that such a measure exists. We have no particular application in mind for this result. The proof, however, is instructive and illustrates certain principles which we shall use later.

Our proof is based on the fact that the commutator subgroup of F_2 is isomorphic to F_∞. For the purposes of this proof let us identify the two, that is, henceforth F_∞ means the commutator subgroup of F_2. Consider the random walk $x_1 x_2 \dots x_n$ on F_2 described in paragraph A. We may inquire how often it meets the subgroup F_∞.

Definition 4.1. Let U_n be a random walk in a space X. A subset $R \subset X$ is called a *recurrence set* for the random walk if, no matter at what point the random walk originates, the probability is 1 that it enters R infinitely often.

Lemma 4.1. F_∞ is a recurrence set for the random walk $gx_1 x_2 \dots x_n$.

Proof. Let $g \to \bar{g}$ denote the natural map of $F_2 \to F_2/F_\infty$. This quotient group is, of course, the free *abelian* group with two generators, i.e., the two-dimensional lattice $\boldsymbol{Z}^2$. The image in $\boldsymbol{Z}^2$ of the random walk on F_2 is a random walk corresponding to the image in $\boldsymbol{Z}^2$ of the measure μ. But the latter assigns to the generators of $\boldsymbol{Z}^2$ and their inverses measure $\frac{1}{4}$, so the image random walk is just the simple random walk on $\boldsymbol{Z}^2$. But the set consisting of 0 is a recurrence set in $\boldsymbol{Z}^2$, since the simple random walk on $\boldsymbol{Z}^2$ is recurrent. Hence the random walk under discussion on F_2 will enter F_∞ infinitely often no matter where it starts, with probability 1. Q.E.D.

The idea underlying our proof that for an appropriate measure μ', (F_∞, μ') has the same Poisson boundary as (F_2, μ) is as follows. We have $F_\infty \subset F_2 \subset F_2 \cup \Omega_2$. Consider the random walk $u_n = x_1 x_2 \ldots x_n$ on F_2. It converges with probability 1 to a point z_1 of Ω_2 which is distributed according to ν. By Lemma 4.1, on its way to the boundary, u_n meets F_∞ infinitely often. The point z_1 is therefore a limit of a sequence in F_∞ as well. Call this sequence $\{u_{n_k}\}$. If we can show that these variables with values in F_∞ correspond to a product of independent F_∞-valued r.v.'s with a fixed distribution μ', then with this μ', it will at least be clear that (Ω_2, ν) is a boundary of (F_∞, μ'). Now how is $u_{n_{k+1}}$ obtained from u_{n_k}? We proceed from u_{n_k} by right multiplication with $x_{n_k+1}, x_{n_k+2}, \ldots$ until the random walk again meets F_∞. But this evidently takes place the first time that the product $x_{n_k+1} \ldots x_{n_k+m}$ belongs to F_∞, since $u_{n_k} \in F_\infty$. This suggests that if we define the measure μ' on F_∞ as representing the distribution of the position of u_n at the first time that u_n enters F_∞, then this measure gives rise to to the random walk u_{n_k} on F_∞. We say 'suggests' because a rigorous proof must take into account the fact that the n_k themselves are random variables. (To be precise, what must be shown is that $x_1 \ldots x_{n_1}$ and $x_{n_1+1} \ldots x_{n_2}$ are still independent variables. This is true, and has something to do with the so-called *strong* Markov property.)

We actually follow a different route. Let μ' denote the measure on F_∞ just described. In other words, for $g \in F_\infty$, $\mu'(g)$ is the probability that the first time the product $x_1 x_2 \ldots x_n$ belongs to F_∞, it have the value g. What we shall show is that the restrictions of μ-harmonic functions on F_2 to F_∞ are μ'-harmonic functions on F_∞, and that all μ'-harmonic functions on F_∞ arise uniquely in this way. It is quite easily seen that this establishes an isomorphism of the two Poisson boundaries.

First a general lemma.

Lemma 4.2. Let G be a discrete group, μ a probability measure on the group, and $\{x_n\}$ a stationary sequence of independent r.v.'s with distribution μ. Let R be a recurrence set in G for the random walk $x_1 x_2 \ldots x_n$ and let $h(g)$ be a μ-harmonic function on G. If θ_g denotes the distribution of the first point of R to be hit by $g x_1 x_2 \ldots x_n$, then

$$h(g) = \sum \theta_g(g^*)h(g^*). \tag{4.3.1}$$

Proof. Start by writing the mean value relation

$$h(g) = \sum \mu(g)h(gg')$$

For each term $h(gg')$ we may independently decide whether we wish to apply the mean value relation to express it in terms of $h(gg'g'')$. Let us base our decision on whether or not the group element gg' is in R. If it is, we leave the expression alone; if not, we apply the mean value relation once more. Continuing in this way we find that after a large number of steps, most of the weight in the right side mean-value

$$h(g) = \sum \theta_g^{(n)}(g^*)h(g^*)$$

is concentrated on values $h(g^*)$ with g^* in R, the combined weight of everything else being the probability that R has not yet been reached in n steps. Moreover, examining the procedure we have followed, we see that $\theta_g^{(n)}(g^*)$ is the probability of the compound event that R has been reached in n steps, and that the point at which it was reached was g^*. Thus in the limit we have, recalling that h is bounded by definition,

$$h(g) = \sum \theta_g(g^*)h(g^*).$$

Return now to the groups F_2 and F_∞ and the measures μ and μ'. We may apply the lemma with F_∞ as the recurrence set. Then $\theta_e = \mu'$, and

$$h(e) = \sum \mu'(g^*)h(g^*)$$

for every μ-harmonic function h. Now if h is μ-harmonic so is its translate $h(g\cdot)$. Hence

$$h(g) = \sum \mu'(g^*)h(gg^*),$$

and so, taking g in F_∞, we see that h is μ'-harmonic. Hence the restrictions to F_∞ of μ-harmonic functions on F_2 are μ'-harmonic.

Moreover this operation of restriction is one-one. For by the lemma it follows quite generally that if the restriction of a μ-harmonic function to a recurrence set vanishes, then the function vanishes identically.

Finally we wish to show that every μ'-harmonic function can be obtained in this way. Suppose then that $h(g)$ is a μ'-harmonic function defined for g in F_∞. Set

$$h'(g) = \sum \theta_g(g^*)h(g^*).$$

Now an easy probability argument shows that each $\theta_g(g^*)$ is a μ-harmonic function of g. Hence h' is a μ-harmonic function. Consider its restriction to F_∞. For $g \in F$, $\theta_g(g^*)$ is, by virtue of the fact that F_∞ is a group, the same as $\theta_e(g^{-1}g^*)$ and so

$$h'(g) = \sum \theta_e(g')h(gg') = h(g).$$

This proves that h is the restriction to F_∞ of a μ-harmonic function, and this gives us the desired result.

4.4. Brownian motion

We now consider random walks on a Lie group G. The reader will be missing nothing if he is familiar with only two groups, $SL(2, \boldsymbol{R})$ and $SL(3, \boldsymbol{R})$, the groups of matrices with real entries and with determinant 1 in dimensions 2 and 3. For the details regarding the discussion here we refer to [11] and [12]. There it is shown that for a semi-simple Lie group and an absolutely continuous probability measure μ, the Poisson boundary can be determined in a more or less explicit fashion. The underlying space of the Poisson boundary is one of a finite family of homogeneous manifolds associated with the group, and the stationary measure on this manifold is absolutely continuous.

For a certain class of μ on G, the situation is simpler still. These measures correspond to what we shall call the Brownian motion on G, and for these, the boundary measure ν can be given explicitly. It is perhaps most natural to define the Brownian motion as a continuous parameter process. We shall content ourselves with a discussion of a discrete time analogue which will serve our purposes just as well.

If G is a semi-simple Lie group, it contains a maximal compact subgroup K and a solvable subgroup S which between them span the group: $G = KS$, i.e., each $g \in G$ admits a decomposition $g = ks$, $k \in K$, $s \in S$. For the matrix groups $G_r = SL(r, \boldsymbol{R})$ (this will be our notation hereafter), we can take K_r to be the subgroup of orthogonal matrices and S_r to be the subgroup of upper triangular $r \times r$ matrices. We include in S_r also matrices with negative terms along the diagonal, so $K_r \cap S_r$ is non-trivial (though finite). The decomposition $G_r = K_r S_r$ is well known. (To express a matrix as a product of this form, apply Gram-Schmidt orthogonalization to the rows of the matrix.)

Definition 4.2. A probability measure μ on a semi-simple Lie group G is *spherical* if it is absolutely continuous with respect to Haar measure on G, and if $k\mu = \mu k = \mu$, for $k \in K$.

An equivalent condition is that

$$\int_G f(k_1, gk_2)\mathrm{d}\mu(g) = \int_G f(g)\mathrm{d}\mu(g)$$

for any $k_1, k_2 \in K$.

Form the homogeneous space $B_r = G_r/S_r$. The elements of B_r are cosets gS_r and G_r operates on B_r by $g(g' S_r) = gg' S_r$. B_r is thus a G_r-space. We will presently give a more explicit description of B_2 and B_3. Observe that K_r acts transitively on B_r, and so there exists a unique K_r-invariant measure on B_r. We denote this measure in each case as m_B. Now for any measure ν on B_r, $\mu * \nu$ is K_r-invariant, and so $\mu * \nu = m_B$. In particular $\mu * m_B = m_B$ and we see that m_B is the unique stationary measure for μ on B_r.

Theorem 4.1. The Poisson boundary of (G_r, μ) is (B_r, m_B) whenever μ is a spherical measure on G_r.

Let us point out one interesting consequence of this theorem. We recall that by the generalized Poisson representation formula, a μ-harmonic function $h(g)$ admits a representation as

$$h(g) = \int \hat{h}(g\xi)\mathrm{d}m_B(\xi).$$

Now notice that the measure μ does not figure on the right hand side of this expression. It follows that the class of μ-harmonic functions is the same for all spherical μ. Hence we shall call these functions simply *harmonic* functions.

We omit the proof of Theorem 4.1 which may be found in [11]. Nevertheless, it will be important to see how the space B_r is to be attached to the group G_r to give convergence for the Brownian motion in G_r to a point of B_r. We consider first the case $r = 2$. Let us identify the space B_2. G_2 acts on two-dimensional Euclidean space and thereby also on the set of lines through the origin, i.e., on the projective space $\boldsymbol{P}^1$. We shall represent the points of $\boldsymbol{R}^2$ as column vectors and the natural map $\boldsymbol{R}^2 \to \boldsymbol{P}^1$ will be written $u \to \bar{u}$. Now G_2 is transitive on $\boldsymbol{P}^1$, and so the latter may be represented as a coset space of G_2. The subgroup in question will be that leaving fixed a preassigned point of $\boldsymbol{P}^1$. Now the subgroup that leaves $\overline{\binom{1}{0}}$ fixed, that is, the matrices having $\binom{1}{0}$ as an eigenvector, are the upper triangular 2×2 matrices. Hence $B_2 = \boldsymbol{P}^1$.

Now recall that B_2, as the space of the Poisson boundary, is to be attached to G_2 in accordance with the measure topology of Definition 3.3. In other words, an element $g \in G_2$ is close to a point $\xi \in B_2$ if the measure gm_B is close to the point measure at ξ. We consider, therefore, the behavior of gm_B for 'large' elements g in the group. Set $g = \begin{pmatrix} a & b \\ c & d \end{pmatrix}$, $ad-bc = 1$. g is large if $a^2+b^2+c^2+d^2$ is large. Then at least one of $\binom{a}{c}$, $\binom{b}{d}$ is large, and if both are, the directions $\overline{\binom{a}{c}}$, $\overline{\binom{b}{d}}$ are close. It follows that for most directions $\bar{u} = \overline{\binom{u_1}{u_2}}$, $g\bar{u}$ is close to the direction of the larger of $\binom{a}{c}$, $\binom{b}{d}$. The exception is when $\binom{u_1}{u_2}$ is approximately orthogonal to $(a\,b)$, $(c\,d)$. We thus find

Lemma 4.3. For given $\varepsilon > 0$, there is a compact set Q_ε in G_2 with the following property. If $g \in Q_\varepsilon$, then there exist intervals I_1 and I_2 in $\boldsymbol{P}^1$ such that both intervals have angular width less than ε and $g(\boldsymbol{P}^1 - I_1) \subset I_2$.

Now let ν be a non-atomic measure on $\boldsymbol{P}^1$. That is to say, ν assigns measure 0 to each point of $\boldsymbol{P}^1$. It follows that every sufficiently small interval carries ν-measure $< \delta$, say. In particular we can choose ε so that $\nu(I_1) < \delta$. Then $\nu(\boldsymbol{P}'-I_1) + 1-\delta$ and since g transfers $\boldsymbol{P}^1 - I_1$ into I_2, $g\nu(I_2) > 1-\delta$. Thus $g\nu$ concentrates most of its mass in a small interval.

Lemma 4.4. As $g \to \infty$, the measures $g\nu$ tend to become point measures.

This is true in particular for the measures gm_B. Using this fact we can show directly that $(\boldsymbol{P}^1, m_B)$ is at least a boundary for (G, μ),

μ a spherical measure. Namely, let $x_1 x_2 \ldots x_n$ denote a Brownian motion in G, that is, the random walk corresponding to a spherical measure μ. To prove that $(\boldsymbol{P}^1, m_B)$ is a boundary we must show that $x_1 x_2 \ldots x_n m_B$ converges, with probability 1, to a point measure. Now by the corollary to Lemma 3.1, the sequence of measures in question does converge to some random limit measure. By the foregoing lemma, this measure will be a point measure unless $x_1 x_2 \ldots x_n$ is bounded for all n. Now quite generally a random product in a topological group remains bounded with positive probability only if the distribution of the variables is confined to a compact subgroup (by the ergodic theorem and the fact that a compact subsemigroup of a group is itself a group). Since our measure μ was chosen absolutely continuous, this clearly cannot happen.

Having ascertained that $(\boldsymbol{P}^1, m_B)$ is a boundary of (G, μ), we can turn matters around and draw conclusions about the random product $x_1 x_2 \ldots x_n$ from our knowledge of the boundary behavior. Namely we know that $x_1 x_2 \ldots x_n m_B$ converges to a point measure δ_{z_1}, where z_k is a μ-process on $\boldsymbol{P}^1$. First of all this tells us that $x_1 x_2 \ldots x_n \to \infty$ (something we knew only for a subsequence). In particular, the random walk is transient! Moreover from the foregoing analysis it is clear that a matrix can be close to the direction z_1 (in the measure topology) only if the direction of the larger of its two column vectors is close to z_1. Thus with probability 1, the larger of the column vectors in the random product must converge, in direction, to z_1. In fact, it may be shown that both columns grow infinitely large. Due to the unimodularity of the x_i, this implies that their directions tend closer to one another. We see then that the 'boundary' behaviour of $x_1 x_2 \ldots x_n$ is described by the statement that as $n \to \infty$, both column vectors converge in direction to a random point of $\boldsymbol{P}^1$.

4.5. Symmetric spaces

The theory of harmonic functions on the group G_2 will appear more familiar if we discuss it not for the space G_2 itself, but for the associated *symmetric space* G_2/K_2. In general, if G is a non-compact semi-simple Liegroup, one associates with it the quotient space modulo a maximal compact subgroup K. This space G/K is a Riemannian space with a number of special properties ([15]). Of these, all that we need to be aware of is that the group G acts on G/K by isometries.

That is, G/K is a metric space and G preserves this metric.

The symmetric space G_2/K_2 may be identifed with the interior of the unit disc: $D = \{z: |z| < 1\}$. To obtain the identification we first let G_2 operate on the upper half complex plane, $bw > 0$, by

$$gw = \frac{aw+b}{cw+d}$$

for $g = \left(\begin{smallmatrix} a & b \\ c & d \end{smallmatrix}\right)$. The subgroup K_2 is seen to be exactly the set of transformations that leave the point i fixed. S_2 consists of the set of g that leave ∞ fixed. In addition, we may check that G_2 is transitive on $bw > 0$. Now let φ be the fractional linear transformation that takes $bw \geqq 0$ onto the unit disc $\bar{D}$ with $\varphi(i) = 0$, $\varphi(\infty) = 1$. Define the action of G_2 on $\bar{D}$ by

$$g(z) = \varphi\left(\frac{a\varphi^{-1}(z)+b}{c\varphi^{-1}(z)+d}\right).$$

K_2 now consists of the transformations that leave 0 fixed, i.e. rotations of D, and S_2 consists of those transformations that leave 1 fixed. Inasmuch as G_2 is transitive on D and on $\partial D = \{z: |z| = 1\}$, we can identify G_2/K_2 with D and G_2/S_2 with ∂D. For the former we identify gK_2 with $g(0)$, and for the latter we identify gS_2 with $g(1)$. So now we have identified all three of the spaces $\boldsymbol{P}^1$, ∂D, $B_2 = G_2/S_2$. Clearly the measure m_B on B_2 corresponds to the unique rotation invariant (Lebesgue) measure on ∂D.

An easy calculation shows that the group invariant metric on D is given by

$$ds = \frac{|dz|}{1-|z|^2}. \tag{4.5.1}$$

There is also a group invariant area on D which is the image under the natural map of Haar measure on G_2. It is given by

$$d\sigma = \frac{dA}{(1-|z|^2)^2}. \tag{4.5.2}$$

Here $|dz|$ and dA refer to the Euclidean length and area. The length and area of (4.5.1) and (4.5.2) will be called the non-Euclidean length and area.

We record for future reference the fact that the non-Euclidean distance of z to the origin is given by

$$\operatorname{dist}(z, 0) = \log \frac{1+|z|}{1-|z|}. \tag{4.5.3}$$

A Brownian motion on the semi-simple group G induces a random walk on G/K which will again be called a Brownian motion. In G the transitions are from g to gg' where g' is distributed according to the measure μ. Since μ is spherical, we see that the probability of transition from g to gg' (actually the *density* of the transition probability) is the same as that from gk to gg', and also the same as from g to $gg'k$ $(k \in K)$. Hence the transition probabilities may be transferred to the quotient space G/K, the transition probability from gK to $gg'K$ being that from g to gg'. In the space D, for example, the induced random walk can be described as follows. For each point z consider the set of circles $C(r, z)$ with non-Euclidean radius r, and non-Euclidean center z. The transitions from z consist of jumps to a point on the circle $C(r, z)$ where r is chosen according to a fixed distribution related to μ, and the point on the circle is chosen with uniform distribution.

This random walk is clearly invariant with respect to the action of G on the symmetric space. That is, the probability of transition from z to z' is the same as that from gz to gz'. This is clear since this was the case for the Brownian motion on G. Moreover a realization of the induced Brownian motion is obtained simply by mapping $x_1 x_2 \ldots x_n \to x_1 x_2 \ldots x_n K$. Thus if $z \in D$, and $W_n(z)$ denotes the position at time n of the Brownian motion in D originating at z, then $W_n(z)$ can be realized as

$$W_n(z) = g x_1 x_2 \ldots x_n(0)$$

where g satisfies $g(0) = z$. Note that the transitions of W_n are *not* obtained by applying the group operations, for G acts to the left on G/K and the transitions occur from the right.

If we carry over the boundary theory for $x_1 x_2 \ldots x_n$ to $W_n(z)$, we obtain the following results. Heretofore, a point $g \in G_2$ was considered close to a point $\xi \in \boldsymbol{P}^1$ if the measure gm_B was close to the point measure δ_ξ on $\boldsymbol{P}^1$. Now suppose $g(0) = z$ with $|z|$ close to 1. Then it is evident that g transfers most of the mass of m_B on ∂D to a point of ∂D close to z. Thus if $z_n \to \zeta \in \partial D$ with $z_n = g_n(0)$, then $g_n m_B \to \delta_\zeta$. This means that the topology in $G_2 \cup B_2$, when transferred to $D \cup \partial D$ corresponds to the ordinary topology of the closed disc. The assertion regarding convergence almost everywhere of the Brownian motion paths of G_2 to points on the boundary now amounts to the assertion that, with probability 1, $W_n(z)$ converges in the closed unit disc to a point of the bounding circle.

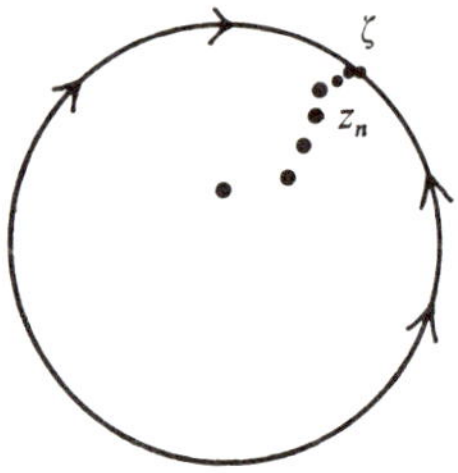

Fig. 4.1.

Now let us turn to harmonic functions. Notice that if $h(g)$ is a harmonic (i.e., μ-harmonic for some and therefore all spherical measures) function on the semi-simple group G, then $h(g)$ depends only on the coset gK. For, the mean-value property says that

$$h(gk) = \int h(gkg')\mathrm{d}\mu(g') = \int h(gg'')\mathrm{d}k\mu(g'')$$
$$= \int h(gg'')\mathrm{d}\mu(g'') = h(g)$$

for $k \in K$, since $k\mu = \mu$.

In view of this we can associate with h a function $\bar{h}$ on G/K which we again call harmonic. Let us see what type of function this is in case $G = G_2$. The Poisson representation formula now reads

$$\bar{h}(g(0)) = h(g) = \int \hat{h}(g\zeta)\mathrm{d}m_B(\zeta) = \int \hat{h}(\zeta)\frac{\mathrm{d}gm_B}{\mathrm{d}m_B}(\zeta)\mathrm{d}m_B(\zeta).$$

If we compute dgm_B/dm_B we find that it is given by

$$\frac{\mathrm{d}gm_B}{\mathrm{d}m_B}(\zeta) = P(g(0), \zeta)$$

where $P(z, \zeta)$ is the classical Poisson kernel from the theory of ordinary harmonic functions in the disc. Substituting we find

$$\bar{h}(z) = \int \hat{h}(\zeta)P(z, \zeta)\mathrm{d}m_B(\zeta)$$

which, of course, implies that the function $\bar{h}$ is harmonic in the classical sense. Thus the two notions of 'harmonic' coincide.

We should also mention another interpretation of the Poisson representation. The value of a harmonic function $\bar{h}(0)$ is the integral of the boundary values with respect to the measure m_B. This measure is the distribution on the boundary of $\lim x_1 x_2 \ldots x_n$, or carrying the theory over to the disc, it is the distribution of $\lim W_n(0)$. Because of the invariance of Brownian motion with respect to the action of the group, it follows that the distribution of $\lim W_n(z)$, with $z = g(0)$, must be given by gm_B. But this is the measure with respect to which one must integrate the boundary values to obtain $h(z)$. Thus, for any $z \in D$, the value of a harmonic function at z is obtained by integrating the boundary values with respect to the 'hitting measure on the boundary starting at z''. (Actually this is nothing more than an application of Proposition 3.4.)

4.6. The Poisson boundary for $SL(3, \mathbf{R})$

The Poisson boundary of $G_3 = SL(3, \boldsymbol{R})$ is, according to Theorem 4.1, the quotient space $B_3 = G_3/S_3$. To obtain a geometric realization of this space, we form the space B' of all pairs of points (u, v) in $\boldsymbol{P}^2$ corresponding to orthogonal directions u and v. The group G_3 acts on $\boldsymbol{P}^2$ in a natural way, and we can make B' into a G_3-space by setting $g(u, v) = (gu, {}^tg^{-1}v)$. ${}^tg$ represents the transpose of the matrix g, and $g \to {}^tg^{-1}$ is an automorphism of G_3. It is clear that the operations defined send orthogonal directions into orthogonal directions. It is also easy to check that the orthogonal group K_3 is transitive on B', and *a fortiori*, G_3 is transitive on this space. We ask for the subgroup of G_3 that leaves fixed the point $(\bar{e}_1, \bar{e}_3)$, where e_1, e_2, e_3 form the standard basis of $\boldsymbol{R}^3$. We see at once that the condition on an element g that it belong to this subgroup is that e_1 be an eigenvector of g and e_3 an eigenvector of ${}^tg$. But this is equivalent to $g \in S_3$, the group of upper triangular matrices. Thus $B' = G_3/S_3 = B_3$ and this gives us the desired geometric realization of B_3.

We turn now to the Brownian motion in G_3 for a spherical measure μ and study the boundary behavior. Reinterpreting this in terms of the matrix product, we find once again that the column vectors of $x_1 x_2 \ldots x_n$ converge with probability 1 in direction to a random point of $\boldsymbol{P}^2$. In particular, if we denote the limiting μ-process variable on B_3 by $\lim x_1 x_2 \ldots x_n = (u_1, v_1)$, then the limiting direction of the column vectors is u_1. To see the significance of v_1, apply the automorphism $g \to {}^t g^{-1}$ to the variables of the random walk. We conclude from the foregoing that the columns of ${}^t x_1^{-1}\, {}^t x_2^{-1} \ldots {}^t x_n^{-1}$ also converge in direction. Denoting the columns of $x_1 x_2 \ldots x_n$ by $r_n, s_n, t_n \in \boldsymbol{R}^3$, we find that the columns of the transpose inverse matrix are $s_n \wedge t_n$, $t_n \wedge r_n$, $r_n \wedge s_n$, where $t \wedge s$ denotes the vector product. Thus not only do r_n, s_n, and t_n tend to a common limit direction as $n \to \infty$, but the planes through the three pairs of these vectors also converge in direction as $n \to \infty$. This is the direction of the plane perpendicular to v_1.

There is an important difference between the boundary behavior of group elements in G_2 and in G_3. In the case of G_2, the boundary space B_2 constituted a compactification of the group. This is a consequence of Lemma 4.4, according to which a sequence going off to infinity in the group necessarily gets closer to the points of B_2. In the case of G_3 this will not always be so (even though it was true *with probability one* for Brownian motion paths tending to infinity). Consider, in fact, the sequence

$$g_n = \begin{pmatrix} n & 0 & 0 \\ 0 & n & 0 \\ 0 & 0 & n^{-2} \end{pmatrix}.$$

We recognize that this sequence has no subsequence converging to a point in B_3 from the fact that the directions of the largest column vectors do not converge together. More specifically, if we consider the measures $g_n m_B$, it is quite easy to see that they do not tend to a point measure, but rather to a measure supported by the circle $\{(u, e_3), u \perp \bar{e}_3\}$. This is one of a class of circles on B_3. These circles are the fibers of the two fibrations of $B_3 \to \boldsymbol{P}^2$, and we shall call them *great circles* in B_3. It is possible to show that there are measures $g m_B$ coming close to measures supported by any great circle.

In any case, $G_3 \cup B_3$ is not compact. This fact will play a significant role in the sequel. It is precisely this difference between G_2 and G_3 that enables us to distinguish between their respective full subgroups.

For completeness let us point out which measures would have to be added to the points of G to compactify this space. It may be shown that any convergent sequence of measures $g_n m_B$ converges either to a point measure, or to a circle measure or to a measure absolutely continuous with respect to m_B. In the latter case the elements $\{g_n\}$ must be bounded in G_3. To give a complete description of the compactification of G_3 by measures on B_3, we need only describe the limit circle measures. Now we can view a great circle as a copy of $\boldsymbol{P}^1$. Let $m_{\boldsymbol{P}^1}$ denote the rotation invariant measure on the circle. Then the remaining measures on the circle are obtained from this one by applying projective transformations of the circle. Thus each great circle carries on it all the measures $G_2 m_{\boldsymbol{P}^1}$ described in connection with the boundary of G_2.

5. Discrete Subgroups of $SL(r, \mathbf{R})$

We have seen that the boundary theory of Brownian motion on a unimodular group is tractable (though by no means complete!) and we shall attempt to apply this theory to discrete subgroups of the G_r. The key to passing from the Lie group case to the discrete case is provided by the following theorem.

Theorem 5.1. If Γ is a full subgroup of G_r, then there exists a probability measure μ on Γ for which $P(\Gamma, \mu) = (B_r, m_B)$.

In other words, if μ_0 is a spherical measure on G_r, then (Γ, μ) and (G_r, μ_0) have the same Poisson boundary. The situation here is reminiscent of the situation in Section 4.3 where we showed that for the simple random walk on F_2 there is a measure on its commutator subgroup F_∞ leading to the same Poisson boundary. The method used there will, however, not apply to our situation, because a countable set cannot be a recurrence set for Brownian motion (which it might very well be if the ambient group is countable).

The proof will be presented in a series of steps. To avoid undue complications we shall prove each statement regarding the group G_r either for the dimension 2 or 3, i.e., for the lowest dimension that is typical. Moreover, we shall generally make the simplifying assumption that Γ is cocompact and not merely full. This will keep our exposition within reasonable bounds.

5.1. Construction of μ

With the group G_r we associate the symmetric space G_r/K_r which, we have pointed out, is a metric space. Using this metric we may define the function $\Lambda(g)$ by

$$\Lambda(g) = \text{dist}\,(gK, K). \tag{5.1.1}$$

For $G_r = G_2$, the case with which we shall have the most to do, this function becomes

$$\Lambda(g) = \log \frac{1+|g(0)|}{1-|g(0)|}.$$

Proposition 5.1. If Γ is a cocompact subgroup of G_r, then there exists a probability measure μ on Γ satisfying

(a) $\mu(\gamma) > 0$ for each $\gamma \in \Gamma$
(b) $\mu * m_B = m_B$
(c) $\Sigma \mu(\gamma)\Lambda(\gamma) < \infty$.

Condition (b) implies that every harmonic function on G_r is also μ-harmonic. In particular, its restriction to Γ is μ-harmonic. Namely we have

$$h(g) = \int_{B_r} \hat{h}(\xi)\mathrm{d}gm_B(\xi)$$

so that

$$\sum \mu(\gamma)h(g\gamma) = \int_{B_r} \hat{h}(\xi)\mathrm{d}g(\sum \mu(\gamma)\gamma m_B)(\xi) = \int_{B_r} \hat{h}(\xi)\mathrm{d}g\mu * m_B(\xi)$$
$$= \int_{B_r} \hat{h}(\xi)\mathrm{d}gm_B(\xi) = h(g).$$

Thus the proposition implies among other things that harmonic functions, say in the disc, satisfy a series of discrete mean-value relations.

We shall give the details for the proof of this proposition for the case $r = 2$. We think of G_2, and therefore also Γ, as acting on the disc D. For the moment ignore the requirements (a) and (c) of the proposition. We concentrate on constructing a measure satisfying (b).

It will be convenient to reinterpret (b) in terms of the Brownian motion in the disc D. The measure gm_B represents, as we know, the hitting probabilities on the boundary ∂D of the Brownian motion originating at $g(0)$. (b) says that tbe hitting probabilities originating at 0 can be expressed as a convex linear combination of those starting from the points $\gamma(0)$. Let us denote by m_z the measure on the boundary ∂D representing the hitting probabilities of the Brownian motion starting at z. Thus $m_{g(0)} = gm_B$.

Lemma 5.1. There exist $L < \infty$ and $\delta > 0$ such that for any $z_0 \in D$

$$m_{z_0} = \sum p_j m_{\gamma_j(0)} + \int m_\xi \, d\lambda(\xi) \tag{5.1.2}$$

with all the ξ and $\gamma_j(0)$ occurring to the right of (5.1.2) being within a (non-Euclidean, throughout this proof) distance L of the point z_0, with the $p_j > 0$, and with λ a non-negative measure with total mass $< 1-\delta$.

Proof. Since Γ is cocompact, we may find a number L so large that a circle of radius L about any point of D contains strictly in its interior at least one point of the form $\gamma(0)$, $\gamma \in \Gamma$. We shall consider the continuous parameter Brownian motion in this discussion (merely for picturesqueness; it doesn't really matter which Brownian motion

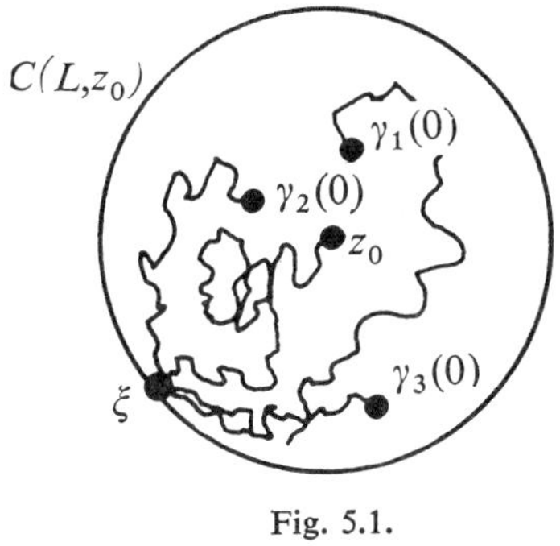

Fig. 5.1.

is used in this argument). Consider now the circle of radius L and center z_0, $C(L, z_0)$, and suppose that $\gamma_j(0)$ are those points of $\Gamma(0)$ which fall within – but not on – $C(L, z_0)$. The hitting measures to ∂D

from a point z inside $C(L, z_0)$ can be computed as follows. We define the hitting distribution

$$m(C(L, z_0), z; J)$$

where J is a subset of the circle $C(L, z_0)$, and z is within the circle, as the probability that the first point of the circle $C(L, z_0)$ to be hit by the Brownian motion starting at z lies in J. (Here we have taken advantage of the continuity of paths of the Brownian motion which assures us that the path from z eventually meets $C(L, z_0)$. For the discrete version we would have to assume that the jumps are of bounded lengths, replacing the circle in this argument by an annulus.) Then the hitting probabilities m_z are obtained by integrating the hitting probabilities m_ξ for points $\xi \in C(L, z_0)$ with respect to the hitting distribution for ξ starting at z:

$$m_z = \int m_\xi m(C(L, z_0), z; \mathrm{d}\xi) \tag{5.1.3}$$

(This uses the strong Markov property for Brownian motion.) For each $\gamma_j(0)$ inside the circle $C(L, z_0)$, choose a positive number p_j, and consider $m_{z_0} - \sum p_j m_{\gamma_j(0)}$. By (5.1.3) this becomes

$$\int m_\xi\{m(C(L, z_0), z_0; \mathrm{d}\xi) - \sum p_j m(C(L, z_0), \gamma_j(0); \mathrm{d}\xi)\}. \tag{5.1.4}$$

Now all the measures $m(C(L, z_0), z; .)$ are absolutely continuous with respect to Lebesgue measure on $C(L, z_0)$ if z is strictly within $C(L, z_0)$. Hence, if the p_j are chosen sufficiently small, the measure appearing in brackets in (5.1.4) will be a positive measure. Just how small the p_j must be chosen depends on how the points $\gamma_j(0)$ are situated inside $C(L, z_0)$. Now if we replace the point z_0 by γz_0, for some $\gamma \in \Gamma$, then we replace the circle $C(L, z_0)$ by $C(L, \gamma z_0) = \gamma C(L, z_0)$. The entire picture is then to be replaced by its translate with γ, the γ_j being replaced by $\gamma\gamma_j$, and it follows that the values of p_j may be retained. Here we use the group invariance of the Brownian motion. Writing $z_0 = g_0(0)$ we see that the choice of the p_j, in particular the value $\sum p_j$, depends on the coset Γg_0. From the compactness of G/Γ it follows that for some $\delta > 0$, the positivity of the measure appearing in (5.1.4) can be arranged for *each* z_0 in D by a choice of p_j satisfying $\sum p_j > \delta$. But we then have an expression of the form (5.1.2) with

$\lambda > 0$ and with $\sum p_j > \delta$. Since all the m_z are probability measures it follows that the total mass of the measure λ is less than $1-\delta$. This proves the lemma.

To construct the measure μ of Proposition 5.1, we apply the lemma successively, first for $z_0 = 0$, i.e., for $m_0 = m_B$, and then for the measures m_ξ appearing under the integral, and we continue 'pushing outward' in this way. At each step of the procedure, more and more of the measure m_B has been converted to a positive linear combination of $m_{\gamma(0)}$. In the limit we obtain an expression for m_B entirely in terms of $m_{\gamma(0)}$.

To be precise, suppose that at the nth stage we have

$$m_0 = \sum p_\gamma^{(n)} m_{\gamma(0)} + \int m_\eta \, d\lambda^{(n)}(\eta) \tag{5.1.5}$$

satisfying the following conditions: $\lambda^{(n)}$ is a positive measure on $\{\xi\colon \text{dist}\,(\xi, 0) \leqq nL\}$, the total mass of $\lambda^{(n)}$ does not exceed $(1-\delta)^n$, and $p_r^{(n)} > 0$ only for points $\gamma(0)$ within a distance nL of 0. If we apply (5.1.2) to each m_η:

$$m_\eta = \sum p_\gamma(\eta) m_{\gamma(0)} + \int m_\xi \, d\lambda_\eta(\xi),$$

then

$$\begin{aligned} m_0 &= \sum \left(p_\gamma^{(n)} + \int p_\gamma(\eta) \, d\lambda^{(n)}(\eta) \right) m_{\gamma(0)} + \iint m_\xi \, d\lambda_\eta(\xi) \, d\lambda^{(n)}(\eta) \\ &= \sum p_\gamma^{(n+1)} m_{\gamma(0)} + \int m_\eta \, d\lambda^{(n+1)}(\eta), \end{aligned}$$

where now $\lambda^{(n+1)}$ is a positive measure on $\{\xi\colon \text{dist}\,(\xi, 0) < (n+1)L\}$, the total mass of $\lambda^{(n+1)}$ does not exceed $\int \lambda_\eta(D) \, d\lambda^{(n)}(\eta) \leqq (1-\delta)^{n+1}$ and $p_r^{(n+1)} > 0$ only for dist $(\gamma(0), 0) \leqq (n+1)L$. It is now clear that by induction, (5.1.5) can be achieved for each n. In addition $p_\gamma^{(n)}$ increases with n for each γ. Setting $\mu(\gamma) = \lim p_\gamma^{(n)}$ we find $m_B = \sum \mu(\gamma) m_{\gamma(0)}$, which is condition (b) of the proposition.

Let us show that for this choice of μ, condition (c) is also fulfilled. From our construction it follows that the $\mu(\gamma)$ for those elements γ for which dist $(\gamma(0), 0) > nL$, i.e., for which $\Lambda(\gamma) > nL$, comes from the m_η to the right of (5.1.5). Hence $\sum_{\Lambda(\gamma) > nL} \mu(\gamma) < (1-\delta)^n$ and $\sum_n \left(\sum_{\Lambda(\gamma) > nL} \mu(\gamma)\right) < \infty$. Now in this sum γ occurs k times if $kL < \Lambda(\gamma) \leqq (k+1)L$. Hence $\sum \mu(\gamma)\Lambda(\gamma) < \infty$.

Finally we turn to condition (a). This condition may not yet be fulfilled. However, if at the nth stage of approximation, we adjoin all the points $\gamma(0)$ at a distance less than nL from 0, assigning to each a sufficiently small probability, then we may proceed with the induction argument as above, at the same time ensuring that all $\gamma(0)$ appear. Thus our proposition is proven.

5.2. (B_r, m_B) IS THE POISSON BOUNDARY OF (Γ, μ)

We first show that (B_r, m_B) is a boundary of (Γ, μ). Since m_B is a stationary measure for μ by Proposition 5.1 (b), it will suffice to show, in accordance with Proposition 3.3, that if $\{y_n\}$ is a stationary sequence of independent r.v.'s with distribution μ, then, with probability 1, $y_1 y_2 \ldots y_n m_B$ converges to a point measure on B_r. For the case $r = 2$, this is readily established, for we know in that case that if the sequence of measures converges at all, and if the group elements y_n are unbounded, then the limit must be a point measure. For this reason our proof that (B_r, m_B) is a boundary of (Γ, μ) will be carried out for the case $r = 3$.

Lemma 5.2. Let μ be a probability measure on G_3 satisfying: (a) μ is not confined to a compact or a reducible subgroup of G_3, (b) $\int \log ||g|| \, ||g^{-1}|| d\mu(g) < \infty$ where $||g||$ represents the norm of g as an operator on $\boldsymbol{R}^3$, and (c) $\mu * m_B = m_B$ where m_B is the rotation invariant measure on B_3. Let $\{y_n\}$ denote a sequence of independent r.v.'s with distribution μ. Then $y_1 y_2 \ldots y_n m_B$ converges to a point measure on B_3, with probability 1.

Proof. Since m_B is a stationary measure for μ, we can apply the corollary to Lemma 3.1 to conclude that the sequence of measures in question does converge. We set

$$z_k = \lim y_k y_{k+1} \ldots y_{k+n} m_B,$$

so that $\{z_k\}$ is a measure-valued sequence of random variables. We wish to show that, with probability 1, the z_k are point measures. We consider the composite sequence $\{y_k, z_k, k = 1, 2, 3 \ldots\}$. This sequence satisfies the conditions: $y_k z_{k+1} = z_k$, y_k is independent of all z_{k+j}. Now there is a technique by means of which we can extend this sequence to indices $k \leqq 0$. (Suppose we simply label y_1, z_1 as y_{-n}, z_{-n}, and similarly shift the labels on all the variables. The portion or the sequence that retains positive indices is, due to the stationarity, isomorphic to what we started with. Thus we can extend our sequence as far as we like in the negative direction. One can show that this in fact defines (probabilistically) a sequence with k ranging over all the integers. Naturally the probability space for the new process will not be the same as the original.) We can therefore write

$$z_{-k} = y_{-k}y_{-k+1}\cdots y_{-1}z_0 \tag{5.2.1}$$

where the variables y_{-i} again form a stationary sequence of independent r.v.'s as before, and where z_0 is independent of all the y_{-i}.

From the discussion in Section 4.6 we recall that the measures z_k, being limits of gm_B must have one of three forms. They are either absolutely continuous with respect to m_B, or they are 'circle' measures, or they are point measures. By ergodicity it is easily seen that only one of these possibilities actually occurs with positive probability. Now the first possibility cannot occur, for this would imply that the sequence $y_1y_2\cdots y_n$ is bounded, which would imply that the y_k are confined to a compact subgroup, contrary to hypothesis. Let us suppose therefore that the z_k are circle measures.

Referring to (5.2.11), we fix the measure z_0 momentarily. For definiteness, suppose it is supported by the circle $\{(u,\bar{e}_3), u \perp e_3\} \subset B_3$. We shall show that z_{-k} then converges, with probability 1, to a point measure. For this it suffices to show that the angle between the columns of $y_{-k}\cdots y_{-1}$ tends to 0 as $k \to \infty$. Suppose then that these columns are u_k, v_k, w_k. We wish to show that $|u_k \wedge v_k|/|u_k|\,|v_k| \to 0$. Notice that $u_k \wedge v_k$ is a column of ${}^t(y_{-k}\cdots y_{-1})^{-1}$. To obtain the desired result we shall show that the columns of the random product $y_{-k}\cdots y_{-1}$ each have an exponential rate of growth:

$$\log|u_k| \sim k\alpha, \qquad \log|v_k| \sim k\alpha.$$

By the same token, we shall find that

$$\log|u_k \wedge v_k| \sim k\beta$$

for some β. If then we show that $\beta < 2\alpha$ it will follow that $|u_k \wedge v_k|/|u_k|\,|v_k| \to 0$ exponentially.

We shall sketch the proofs of the first statements. An exact proof will be found in [12, (Theorem 8.5)]. Let us write

$$\log|u_k| = \sum \log \frac{|y_{-i-1}y_{-i}\cdots y_{-1}e_1|}{|y_{-i}\cdots y_{-1}e_1|} = \sum \log \frac{|y_{-i-1}t_i|}{|t_i|}$$
$$= \sum \rho(y_{-i-1}, \bar{t}_i)$$

where $\bar{t}_i \in \boldsymbol{P}^2$ and $\rho(g, \bar{t}) = \log|gt|/|t|$. The sequence $(y_{-i-1}, \bar{t}_i)$ is not stationary, but it almost is. This suggests applying the ergodic theorem which gives

$$\frac{1}{k}\log|u_k| \to \alpha, \qquad \frac{1}{k}\log|v_k| \to \alpha$$

with

$$\alpha = \iint \rho(g, \bar{t})\mathrm{d}\mu(g)\mathrm{d}m'(\bar{t}) \tag{5.2.2}$$

the $\bar{t}_i$ having distribution m', the latter being the rotation invariant measure on $\boldsymbol{P}^2$. It is clear that to apply the ergodic theorem we must verify that $\rho(g,\bar{t})$ is integrable. But $|\rho(g,\bar{t})| \leqq \max(\log\|g\|, \log\|g^{-1}\|)$, so that the required integrability is a consequence of hypothesis (b) of our lemma. In a similar manner we find

$$\frac{1}{k}\log|u_k \wedge v_k| \to \beta$$

with

$$\beta = \iint \rho'(g, \bar{t})\mathrm{d}\mu(g)\mathrm{d}m'(\bar{t}) \tag{5.2.3}$$

$\rho'(g, \bar{t})$ being $\log|{}^tg^{-1}t|/|t|$. To show that $\beta < 2\alpha$ we transfer the integrals of both (5.2.2) and (5.2.3) to the space B_3 instead of the projective space $\boldsymbol{P}^2$. We can then combine the two integrals to obtain

$$2\alpha - \beta = \iint \log\frac{|gu||gv|}{|gu \wedge gv|}\,\mathrm{d}m_B(u, v)\mathrm{d}\mu(g). \tag{5.2.4}$$

Now (u, v) range over orthogonal pairs of directions in $\boldsymbol{P}^2$. Since μ is not confined to the orthogonal group there will be a set of g with positive μ-measure for which the integrand in (5.2.4) is strictly positive. Since the integrand is non-negative, we conclude $\beta < 2\alpha$.

Returning now to the expression in (5.2.1), we find that z_{-k} must tend to a point measure, with probability 1 for each choice of z_0. The y_{-k} being independent of z_0 this implies that, with probability 1, z_{-k} actually does tend to a point measure. But the z_{-k} form a stationary sequence, and this shows that the measures z_{-k} were, to begin with, point measures.

Proposition 5.2. If μ is the measure constructed for Γ in Proposition 5.1, then (B_r, m_B) is a boundary of (Γ, μ).

The foregoing lemma enables us to prove this proposition for the case $r = 3$. The measure μ, having as support all of Γ, is not confined to a compact or reducible subgroup unless Γ is compact or reducible. If Γ is a full subgroup it is obviously not compact, and by [1] it cannot be reducible (its algebraic closure is G_3!). To complete the proof we merely have to verify the integrability condition (b) of the lemma. Now this condition follows from the integrability of $\Lambda(\gamma)$ since it is not difficult to show that

$$\log ||g|| = O(\Lambda(g)), \qquad \log ||g^{-1}|| = O(\Lambda(g)).$$

We turn now to the more difficult question of showing that (B_r, m_B) is actually the Poisson boundary. We shall carry out the proof for the case $r = 2$ which, from this point on, is typical. We begin with two lemmas.

Lemma 5.3. Let ξ_1, ξ_2 denote two distinct points of B_2. Let $N(L)$ denote the number of elements in Γ satisfying

$$\log \left\{\frac{d\gamma^{-1} m_B}{dm_B}(\xi_i)\right\}^{-1} < L \tag{5.2.5}$$

for both ξ_1 and ξ_2. Then $\log N(L) \sim L$.

Proof. We identify B_2 with ∂D so that Γ acts on B_2 by fractional linear transformations. One then has

$$\frac{dg^{-1}m_B}{dm_B}(\zeta) = |g'(\zeta)|.$$

Set $g(z) = \rho(z+\alpha)/(1+\alpha z)$. Then

$$|g'(\zeta)| \doteq \frac{1-|\alpha|^2}{|1+\bar{\alpha}\zeta|^2} = \frac{1-|g(0)|^2}{|1+\bar{\alpha}\zeta|^2}$$

Since ξ_1, ξ_2 are distinct

$$a_1(1-|g(0)|^2) < \min(|g'(\xi_1)|, (|g'\xi_2)|) < a_2(1-|g(0)|^2)$$

or, since $\Lambda(g) = \log(1+|g(0)/(1-|g(0)|)$,

$$b_1+\Lambda(\gamma) < \max_{i=1,2}\left(\log\left\{\frac{\mathrm{d}\gamma^{-1}m_B}{\mathrm{d}m_B}(\xi_i)\right\}^{-1}\right) < b_2+\Lambda(\gamma)$$

To prove the lemma, it therefore suffices to prove that if $N_1(L)$ denotes the number of elements γ with $\gamma(0)$ no further than L away from 0, then $N_1(L) \sim L$. Now the set of γ with $\gamma(0) = 0$ is finite, so we may replace $N_1(L)$ here by the number $N_2(L)$ of $\gamma(0)$ that are within distance L of 0. Now this is a purely geometric question. The proof is particularly simple if we assume that the fundamental domain of Γ is compact. For then, surrounding 0 with a sufficiently large compact region Q, the translates ΓQ cover D and it follows that $N_2(L) > c_1 A(L)$, where $A(L)$ denotes the area of the circle with center 0 and (non-Euclidean) radius L. On the other hand, taking Q sufficiently small we find that $N_2(L) < c_2 A(L+1)$. But

$$A(L) = \int_0^{2\pi}\int_0^{r_L}\frac{r\,\mathrm{d}r\,\mathrm{d}\theta}{(1-r^2)^2} = 2\pi\left(\frac{1}{1-r^2}-1\right)$$

where r_L is the Euclidean radius corresponding to L: $\log(1+r_L/1-r_L) = L$. So $\log A(L) \sim L$ and this proves the lemma.

Now let M be any Γ-space which carries a stationary measure θ for μ: $\mu * \theta = \theta$. Since we assume that the support of μ is exactly Γ, it follows that each $\gamma\theta$ is absolutely continuous with respect to θ, and moreover $\mathrm{d}\gamma\theta/\mathrm{d}\theta$ is bounded. For

$$\theta = \sum \mu(\gamma)\gamma\theta$$

and $\gamma\theta < \mu(\gamma)^{-1}\theta$.

Lemma 5.4. Let $\{y_n\}$ denote a stationary sequence of independent r.v.'s with distribution μ, and suppose that Δ is a recurrence set in Γ for the random walk $y_1 y_2 \ldots y_n$. Then

$$\sum_{\gamma\in\Delta}\left\|\frac{\mathrm{d}\gamma\theta}{\mathrm{d}\theta}\right\|_\infty^{-1} = \infty. \tag{5.2.6}$$

Proof. Assume to the contrary that the sum in (5.2.6) is convergent. Removing finitely many elements from Δ we obtain a set – call it again Δ – which is still a recurrence set, but for which

$$\sum_{\gamma\in\Delta}\left\|\frac{\mathrm{d}\gamma\theta}{\mathrm{d}\theta}\right\|_\infty^{-1} < 1. \tag{5.2.7}$$

Now according to Lemma 4.2, if h is a μ-harmonic function, the value of h at any point, say e, can be expressed in terms of its values on any recurrence set

$$h(e) = \sum_{\gamma \in \Delta} p(\gamma)h(\gamma)$$

this is true for bounded μ-harmonic functions (we took boundedness as part of the definition). For positive unbounded harmonic functions, the same argument shows that one has at least an inequality

$$h(e) \geqq \sum_{\gamma \in \Delta} p(\gamma)h(\gamma).$$

Now $\theta = \sum \mu(\gamma_1)\gamma_1\,\theta$ and $\gamma\theta = \sum \mu(\gamma_1)\gamma\gamma_1\,\theta$ so that

$$\frac{\mathrm{d}\gamma\theta}{\mathrm{d}\theta}(\xi) = \sum \mu(\gamma_1)\frac{\mathrm{d}\gamma\gamma_1\theta}{\mathrm{d}\theta}(\xi)$$

holds almost everywhere in $L^1(M, \theta)$. This means that for almost every ξ, $\mathrm{d}\gamma\theta/\mathrm{d}\theta(\xi)$ is a μ-harmonic function of γ. Hence

$$1 \geqq \sum p(\gamma)\frac{\mathrm{d}\gamma\theta}{\mathrm{d}\theta}(\xi) \qquad \text{a.e.}$$

and, in particular, $p(\gamma)\|\mathrm{d}\gamma\theta/\mathrm{d}\theta\|_\infty \leqq 1$. But then (5.2.7) implies $\sum p(\gamma) < 1$ which cannot be, since Δ must be hit.

We use these lemmas now to show that (B_2, m_B) is the Poisson boundary of (Γ, μ). Suppose therefore that (B, θ) is (a version of) the actual Poisson boundary. Since (B_2, m_B) is at least a boundary, it is an equivariant image of (B, θ) (Theorem 3.1). Let us write $\xi = \rho(\eta)$ for $\xi \in B_2$ and $\eta \in B$. $\rho(\eta)$ is a measurable equivariant map, so we have $\rho(\gamma\eta) = \gamma\rho(\eta)$. The correspondence established by ρ is such that if $\psi \in L^\infty(B_2)$ then $\psi(\xi)$ on B_2 and $\psi(\rho(\eta))$ on B correspond to the same μ-harmonic function. That is

$$\int \psi(\xi)\mathrm{d}\gamma m_B(\xi) = \int \psi(\rho(\eta))\mathrm{d}\gamma\theta(\eta).$$

We can write this also as follows

$$\int \psi(\xi)\frac{\mathrm{d}\gamma m_B}{\mathrm{d}m_B}(\xi)\mathrm{d}m_B(\xi) = \int \psi(\rho(\eta))\frac{\mathrm{d}\gamma\theta}{\mathrm{d}\theta}(\eta)\mathrm{d}\theta(\eta)$$

Since $\rho\theta = m_B$ we have $\int f(\xi)\mathrm{d}m_B(\xi) = \int f(\xi)\mathrm{d}\rho\theta(\xi) = \int f(\rho(\eta))\mathrm{d}\theta(\eta)$

and so

$$\int \psi(\rho(\eta)) \frac{\mathrm{d}\gamma m_B}{\mathrm{d}m_B}(\rho(\eta))\mathrm{d}\theta(\eta) = \int \psi(\rho(\eta)) \frac{\mathrm{d}\gamma\theta}{\mathrm{d}\theta}(\eta)\mathrm{d}\theta(\eta).$$

This identity extends to functions ψ in $L^2(B_2, m_B)$ and we see that for the closed subspace V in $L^2(B, \theta)$ consisting of functions of the form $\psi' = \psi \circ \rho$, we have (in terms of the inner product $\langle \, , \rangle$ of $L^2(B, \theta)$),

$$\left\langle \psi', \frac{\mathrm{d}\gamma m_B}{\mathrm{d}m_B} \circ \rho \right\rangle = \left\langle \psi', \frac{\mathrm{d}\gamma\theta}{\mathrm{d}\theta} \right\rangle$$

If we denote by P_V the Hilbert space projection from $L^2(B, \theta)$ to V, then it follows from the foregoing that

$$P_V\left(\frac{\mathrm{d}\gamma\theta}{\mathrm{d}\theta}\right) = \frac{\mathrm{d}\gamma m_B}{\mathrm{d}m_B} \circ \rho.$$

As a result of the fact that V consists of all the functions in $L^2(B, \theta)$ which are measurable with respect to a certain field of sets (those of the form $\rho^{-1}(A)$, $A \subset B_2$) the projection P_V has certain special properties. (See [5]). In particular one can show that (Jensen's inequality!)

$$P_V(\log f) \leqq \log P_V(f) \tag{5.2.8}$$

with equality holding only for functions $f \in V$.

Lemma 5.5 For every $\gamma \in \Gamma$,

$$\int \log \frac{\mathrm{d}\gamma m_B}{\mathrm{d}m_B}(\xi)\mathrm{d}m_B(\xi) \geqq \int \log \frac{\mathrm{d}\gamma\theta}{\mathrm{d}\theta}(\eta)\mathrm{d}\theta(\eta) \tag{5.2.9}$$

and equality can hold only if the function $\mathrm{d}\gamma\theta/\mathrm{d}\theta(\eta)$ belongs to the subspace V.

Proof. For setting $f = \mathrm{d}\gamma\theta/\mathrm{d}\theta$, the integral to the left in (5.2.9) is $\langle \log P_V(f), 1\rangle$ and the integral to the right is $\langle \log f, P_V 1\rangle = \langle P_V \log f, 1\rangle$. The result then follows from (5.2.8).

Suppose tentatively that we have shown that $\mathrm{d}\gamma\theta/\mathrm{d}\theta$ belongs to V for each $\gamma \in \Gamma$. We shall show that this implies that the map ρ is an isomorphism, so that B_2 can serve as Poisson boundary. Namely, for each $\varphi \in L^\infty(B, \theta)$, the associated μ-harmonic function is given by

$$h(\gamma) = \left\langle \varphi, \frac{d\gamma\theta}{d\theta} \right\rangle = \left\langle \varphi, P_V \frac{d\gamma\theta}{d\theta} \right\rangle = \left\langle P_V \varphi, \frac{d\gamma\theta}{d\theta} \right\rangle$$

which is the μ-harmonic function associated with $P_V\varphi$. But the boundary function for a given μ-harmonic function is uniquely determined on the Poisson boundary, and so $\varphi = P_V\varphi$. This being the case for each φ in $L^\infty(B, \theta)$, we find that P_V is the identity. This, however, shows that the map ρ establishes an isomorphism of (B, θ) and (B_2, m_B).

We have thereby proven the following:

Lemma 5.6. If for every $\gamma \in \Gamma$, (5.2.9) holds with equality, then (B_2, m_B) is the Poisson boundary of (Γ, μ).

The remainder of this section will be devoted to proving that

$$\sum \mu(\gamma) \int \log \left(\frac{d\gamma^{-1} m_B}{dm_B} (\xi) \right)^{-1} dm_B(\xi)$$
$$\geqq \sum \mu(\gamma) \int \log \left(\frac{d\gamma^{-1}\theta}{d\theta} (\eta) \right)^{-1} d\theta(\eta) \quad (5.2.10)$$

This gives

$$\sum \mu(\gamma) \int \log \frac{d\gamma^{-1} m_B}{dm_B} (\xi) dm_B(\xi) \leqq \sum \mu(\gamma) \int \log \frac{d\gamma^{-1}\theta}{d\theta} (\eta) d\theta(\eta)$$

and since all the $\mu(\gamma)$ are strictly positive, this implies that equality must, in fact, take place in (5.2.9) for every γ. This gives the desired result.

Consider now the μ-processes on B and on B_2. We take these moving forward in time rather than backward, so that if $\{z_n\}$ denotes the B-valued process, then

$$z_{n+1} = y_{n+1} z_n$$

with y_{n+1} independent of $z_n, z_{n-1}, \ldots$. Here $\{y_n\}$ is the stationary sequence of independent r.v.'s with distribution μ. We set $\rho(z_n) = w_n$ so that

$$w_{n+1} = y_{n+1} w_n$$

is the μ-process on B_2.

We wish to compare

$$\log \left(\frac{d\gamma^{-1}\theta}{d\theta} (\eta) \right)^{-1} \quad \text{and} \quad \log \left(\frac{d\gamma^{-1} m_B}{dm_B} (\xi) \right)^{-1}$$

for $\gamma = y_n y_{n-1} \ldots y_1$. Notice that we are now considering a random walk on Γ obtained by left multiplication. We have

$$\log \left\{ \frac{\mathrm{d}(y_n \ldots y_1)^{-1} m_B}{\mathrm{d}m_B}(w_0) \right\}^{-1} = \sum_1^n \log \left\{ \frac{\mathrm{d}y_1^{-1} \ldots y_i^{-1} m_B}{\mathrm{d}y_1^{-1} \ldots y_{i-1}^{-1} m_B}(w_0) \right\}^{-1}$$

$$= \sum_1^n \log \left\{ \frac{\mathrm{d}y_i^{-1} m_B}{\mathrm{d}m_B}(y_{i-1} \ldots y_1 w_0) \right\}^{-1} = \sum_1^n \log \left\{ \frac{\mathrm{d}y_i^{-1} m_B}{\mathrm{d}m_B}(w_{i-1}) \right\}^{-1} \tag{5.2.13}$$

We would like to apply the ergodic theorem to this expression. Let us show that

$$E\left(\left| \log \left\{ \frac{\mathrm{d}y_i^{-1} m_B}{\mathrm{d}m_B}(w_{i-1}) \right\}^{-1} \right| \right) < \infty$$

This integral is dominated, however, by

$$E\left(\max_{\zeta} \log \frac{\mathrm{d}y_i^{-1} m_B}{\mathrm{d}m_B}(\zeta) \right)$$

As in the proof of Lemma 5.3, we observe that the Radon-Nikodym derivative in question is related to the function $\Lambda(g)$. To be precise, we may show that

$$\max_{\zeta} \left(\log \frac{\mathrm{d}g^{-1} m_B}{\mathrm{d}m_B}(\zeta) \right) \leqq \log 4 + \Lambda(g)$$

Hence the desired integrability follows from $E(\Lambda(y_i)) < \infty$, which is (c) of Proposition 5.1.

Using the ergodic theorem together with the fact that y_i and w_{i-1} are independent, we conclude that

$$\log \left\{ \frac{\mathrm{d}(y_n \ldots y_1)^{-1} m_B}{\mathrm{d}m_B}(w_0) \right\}^{-1} \sim n\alpha \tag{5.2.14}$$

with

$$\alpha = \sum \mu(\gamma) \int \log \left(\frac{\mathrm{d}\gamma^{-1} m_B}{\mathrm{d}m_B}(\xi) \right)^{-1} \mathrm{d}m_B(\xi) \tag{5.2.15}$$

Next we would like to find a similar expression involving θ instead of m_B. Here we run into the difficulty that

$$\left| \log \left(\frac{\mathrm{d}\gamma^{-1}\theta}{\mathrm{d}\theta}(\eta) \right)^{-1} \right|$$

may not be integrable with respect to μ. We must therefore use a simple modification of the ergodic theorem. Namely, that theorem applies as well to functions that are bounded on one side by an integrable function. (An easy exercise!) Now it is quite easy to see that the negative part of

$$\log\left(\frac{\mathrm{d}\gamma^{-1}\theta}{\mathrm{d}\theta}(\eta)\right)^{-1}$$

is integrable with respect to $\mu \times \theta$. For, just integrating with respect to θ gives

$$\int \log^{+} \frac{\mathrm{d}\gamma^{-1}\theta}{\mathrm{d}\theta}(\eta)\mathrm{d}\theta(\eta) \leqq \int \frac{\mathrm{d}\gamma^{-1}\theta}{\mathrm{d}\theta}(\eta)\mathrm{d}\theta(\eta) = 1$$

As before, we now have

$$\log\left\{\frac{\mathrm{d}(y_n \ldots y_1)^{-1}\theta}{\mathrm{d}\theta}(z_0)\right\}^{-1} \sim n\beta \tag{5.2.16}$$

with

$$\beta = \sum \mu(\gamma) \int \log\left(\frac{\mathrm{d}\gamma^{-1}\theta}{\mathrm{d}\theta}(\xi)\right)^{-1} \mathrm{d}\theta(\xi) \leqq \infty \tag{5.2.17}$$

Combining (5.2.14) and (5.2.16), we obtain

$$\log\left\{\frac{\mathrm{d}(y_n \ldots y_1)^{-1}\theta}{\mathrm{d}\theta}(z_0)\right\}^{-1} \Big/ \log\left\{\frac{\mathrm{d}(y_n \ldots y_1)^{-1}m_B}{\mathrm{d}m_B}(w_0)\right\}^{-1} \to \frac{\beta}{\alpha} \leqq \infty.$$

We rewrite the numerator here using (3.1.3):

$$\log\left\{\frac{\mathrm{d}\theta}{\mathrm{d}(y_n \ldots y_1)\theta}(y_n \ldots y_1 z_0)\right\}^{-1} = \log \frac{\mathrm{d}(y_n \ldots y_1)\theta}{\mathrm{d}\theta}(z_n)$$

and so obtain

$$\liminf \frac{\log\left\|\dfrac{\mathrm{d}(y_n \ldots y_1)\theta}{\mathrm{d}\theta}\right\|_{\infty}}{\log\left\{\dfrac{\mathrm{d}(y_n \ldots y_1)^{-1}m_B}{\mathrm{d}m_B}w_0\right\}^{-1}} \geqq \frac{\beta}{\alpha}, \text{ with probability 1.} \tag{5.2.18}$$

Now the y_k are all independent of w_0, so for almost all w_0, (5.2.18) must be valid for almost all random paths $\{y_n\}$. In particular, there exist two values ξ_1, ξ_2 in B_2 such that, setting $w_0 = \xi_1, \xi_2$, (5.2.18) is valid with probability 1. Consider now the random variables

$$u_n^i = \frac{\log \left\| \dfrac{d(y_n \dots y_1)\theta}{d\theta} \right\|_\infty}{\log \left\{ \dfrac{d(y_n \dots y_1)^{-1} m_B}{dm_B}(\xi_i) \right\}^{-1}}$$

$$v_n^i = \frac{\log \left\| \dfrac{d(y_1 \dots y_n)\theta}{d\theta} \right\|_\infty}{\log \left\{ \dfrac{d(y_1 \dots y_n)^{-1} m_B}{dm_B}(\xi_i) \right\}^{-1}}.$$

Clearly u_n^i and v_n^i have the same distribution. Since $\liminf u_n^i \geqq \beta/\alpha$ with probability 1, it follows that the probability of $\{u_n^i < (\beta/\alpha)-\varepsilon\}$ tends to 0 for each $\varepsilon > 0$. This last assertion will also be valid for v_n^i. We choose a sequence of n_k such that the probabilities in question go to 0 so rapidly that their sum is finite. Then we will have $\liminf v_{n_k}^i \geqq (\beta/\alpha)-\varepsilon$. This implies that the set Δ of γ for which

$$\log \left\| \frac{d\gamma\theta}{d\theta} \right\|_\infty \Big/ \log \left\{ \frac{d\gamma^{-1} m_B}{dm_B}(\xi_i) \right\}^{-1} \geqq \frac{\beta}{\alpha} - 2\varepsilon, \qquad i = 1, 2,$$

is a recurrence set, inasmuch as the random walk $y_1 y_2 \dots y_n$ enters it with probability 1 infinitely often. Now suppose $\beta > \alpha$. Then we could choose ε so that $\beta/\alpha - 2\varepsilon > \tau > 1$. Then for $\gamma \in \Delta$,

$$\log \left\| \frac{d\gamma\theta}{d\theta} \right\|_\infty \geqq \tau \log \left\{ \frac{d\gamma^{-1} m_B}{dm_B}(\xi_i) \right\}^{-1}, \quad i = 1, 2. \qquad (5.2.19)$$

Apply Lemma 5.3 to the right side of (5.2.19). Then the number of $\gamma \in \Delta$ for which

$$\left\| \frac{d\gamma\theta}{d\theta} \right\|_\infty < e^{\tau L}$$

cannot exceed the number of γ in Γ for which both of

$$\log \left\{ \frac{d\gamma^{-1} m_B}{dm_B}(\xi_i) \right\}^{-1}$$

are less than L, and this is $e^{L+0(L)}$. Thus

$$\sum \left\| \frac{d\gamma\theta}{d\theta} \right\|_{\infty}^{-1}$$

summed over those $\gamma \in \Delta$ for which the summand falls in the interval $(e^{-\tau(L+1)}, e^{-\tau L})$ cannot exceed $e^{L+(0)L}\, e^{-\tau L} = e^{-(\tau-1)L+0(L)}$. It follows that

$$\sum_{\gamma \in \Delta} \left\| \frac{d\gamma\theta}{d\theta} \right\|_{\infty}^{-1} < \infty,$$

and inasmuch as Δ is a recurrence set, this contradicts Lemma 5.4. Hence we are forced to the conclusion that $\beta \leqq \alpha$. But this is the inequality of (5.2.10) which we saw leads to the conclusion that (B_2, m_B) is the Poisson boundary of (Γ, μ).

6. Envelopes of Γ

In this section we shall see that if Γ is a full subgroup of $G_r = SL(r, \boldsymbol{R})$, $r > 2$, then it cannot be isomorphic to a full subgroup of G_2. Once again we shall carry out the details only for $r = 3$ which is typical of all the cases. In addition, we shall make the simplifying assumption that Γ is cocompact in G_3. What can actually be proven is the following:

Theorem 6.1 A full discrete subgroup of G_r cannot be isomorphic to a subgroup of G_2 (discrete or not).

In other words, a full discrete subgroup of G_r, $r \geqq 3$, does not have a faithful representation by 2×2 matrices. On the other hand, the free group F_2 which is full in G_2, is isomorphic to a dense subgroup of G_r for any r. A reasonable conjecture seems to be that in general, a discrete full subgroup of G_r cannot be isomorphic to a subgroup of G_s for $s < r$.

Suppose that Γ is isomorphic to a subgroup of G_2 and also of G_3. We shall refer to both of these subgroups as Γ. According to Theorem 5.1, there exists on Γ a measure μ such that the Poisson boundary $P(\Gamma, \mu) = (B_3, m_B)$. Now consider Γ as a subgroup of G_2. It cannot be contained in a compact subgroup, for then Γ would be abelian. That cannot be the case for a number of reasons. One reason is that an abelian group always has a trivial Poisson boundary

(Section 4). Now let us show that μ, as a measure on G_2, possesses a stationary measure θ on B_2. Namely, let θ_1 be any probability measure on B_2. Form the sequence

$$\lambda_n = (n+1)^{-1}(\theta_1 + \mu * \theta_1 + \ldots + \overbrace{\mu * \ldots \mu}^{n} * \theta_1).$$

A subsequence of $\{\lambda_n\}$ must converge because B_2, and therefore $\mathcal{M}(B_2)$ is compact. The limit θ clearly satisfies $\mu * \theta = \theta$. Now let $\{y_n\}$ denote the stationary sequence of independent r.v.'s with distribution μ. Clearly, $y_1 y_2 \ldots y_n$ is unbounded with probability 1, for the y_n themselves are unbounded. (We don't need that; it is enough to know that the y_n are not confined to a compact subgroup.) By the corollary to Lemma 3.1, $y_1 y_2 \ldots y_n \theta$ converges on B_2. Due to the unboundedness of the group elements $y_1 y_2 \ldots y_n$, it follows (this is the special property of G_2!) that the limit measure must be a point measure. By Proposition 3.3, we conclude that (B_2, θ) is a boundary of (Γ, μ). But (B_3, m_B) is the Poisson boundary for this pair; it follows that (B_2, θ) is an equivariant image of (B_3, m_B). This seems implausible, and in the remainder of the section we shall show that, in fact, it cannot happen.

Let ρ denote the equivariant map that takes (B_3, m_B) onto (B, θ). We have $\rho(m_B) = \theta$. Let us show that θ is non-atomic, i.e., that θ assigns measure 0 to sets consisting of a single point. Suppose to the contrary that $\theta(\zeta) > 0$; let $\Delta = \rho^{-1}(\zeta)$. Then $m_B(\Delta) > 0$. Consider the sets $\gamma\Delta$, $\gamma \in \Gamma$. If whenever $\gamma\Delta \cap \Delta$ has positive measure, we have $\gamma\Delta = \Delta$, then Γ cannot act ergodically on $B_3 \times B_3$, for the various $\gamma\Delta$ do not 'mix'. However by [19], a full subgroup of G_3 must be ergodic on this space. On the other hand, if $\gamma\Delta$ intersects Δ

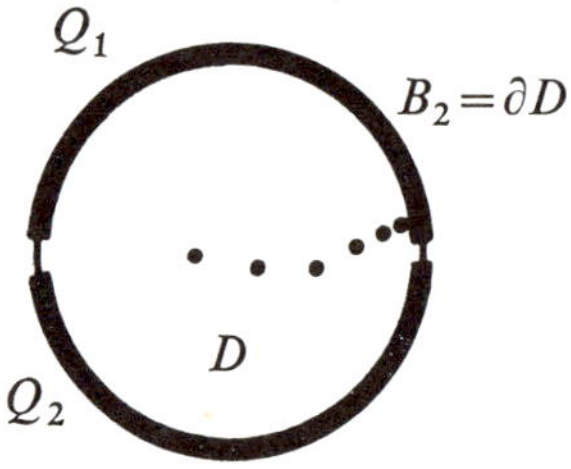

Fig. 6.1.

their images in B_2 intersect, and therefore coincide; therefore $\gamma\Delta = \Delta$. This proves that θ is non-atomic.

Now let Q_1 and Q_2 denote closed disjoint sets of B_2. Suppose $\{\gamma_n\}$ is any sequence of Γ tending to ∞. We claim that

$$\min(\gamma_n\theta(Q_1), \gamma_n\theta(Q_2)) \to 0$$

as $n \to \infty$. This is a consequence of Lemma 4.3. For, eventually $\gamma_n\theta$ concentrates all but ε of its mass in an interval whose width is less than the separation between Q_1 and Q_2. Then either $\gamma_n\theta(Q_1) \leqq \varepsilon$ or $\gamma_n\theta(Q_2) \leqq \varepsilon$. Consider next a pair of measurable functions ψ_1, ψ_2, with supports in Q_1 and Q_2 respectively, and satisfying $0 \leqq \psi_1$, $\psi_2 \leqq 1$. We form the corresponding μ-harmonic functions

$$h_i(\gamma) = \int \psi_i(\zeta) d\gamma\theta(\zeta)$$

It follows from the foregoing that

$$\min(h_1(\gamma), h_2(\gamma)) \to 0 \quad \text{as} \quad \gamma \to \infty. \tag{6.1}$$

Let us transfer this information to G_3. That is, we consider h_1 and h_2 as functions on $\Gamma \subset G_3$. Let Ψ_1, Ψ_2 be the boundary functions of h_1 and h_2 on B_3 (the Poisson boundary space). Then the functions h_i are related to these boundary functions by

$$h_i(\gamma) = \int \Psi_i(\xi) dg m_B(\xi).$$

Replacing the discrete variable γ by $g \in G_3$, we see that h_1 and h_2 extend to harmonic functions on G_3. We shall be interested in the boundary behavior of these harmonic functions.

We wish to compare the value of a harmonic function at a point γ and a neighboring point γg. We have, by (3.3),

$$\frac{d\gamma g m_B}{dm_B}(\xi) \Big/ \frac{d\gamma m_B}{dm_B}(\xi) = \frac{d\gamma g m_B}{d\gamma m_B}(\xi) = \frac{dg m_B}{dm_B}(\gamma^{-1}\xi)$$

so that if g is confined to a bounded set $A \subset G_3$ then the ratio between

$$\frac{d\gamma g m_B}{dm_B}(\xi) \quad \text{and} \quad \frac{d\gamma m_B}{dm_B}(\xi)$$

is bounded above and below. Now let $h(g)$ be a positive harmonic function. Expressing h in terms of its boundary values we see that if g is restricted to A then the ratio between $h(\gamma)$ and $h(\gamma g)$ is bounded above and below, uniformly for all γ. If we now use the cocompactness hypothesis on Γ which implies that for an appropriate bounded set A in G_3, every element of G_3 has the form γg, $g \in A$, we find that for each $g \subset G_3$, there exists $\gamma \in \Gamma$ such that, for a fixed constant c,

$$h(g) < ch(\gamma).$$

Moreover, if $g \to \infty$, we may assume $\gamma \to \infty$. By (6.1) we now have

$$\min(h_1(g), h_2(g)) \to 0 \tag{6.2}$$

for any $g \to \infty$ in G_3. In other words, as g approaches infinity, one or the other of h_1 and h_2 tends to 0. We shall let $g \to \infty$ without getting closer to B_3 – something that could not be done in G_2 – and study the implications. Namely, we let g grow in such a way that gm_B converges to a 'circle' measure on B_3. We shall see that there is a 'boundary' theory for this anomalous convergence as well.

Consider the sequences

$$g_n' = \begin{pmatrix} n & 0 & 0 \\ 0 & n & 0 \\ 0 & 0 & n^{-2} \end{pmatrix} \qquad g_n'' = \begin{pmatrix} n^2 & 0 & 0 \\ 0 & n^{-1} & 0 \\ 0 & 0 & n^{-1} \end{pmatrix}.$$

The limits of the sequences $g_n' m_B$, $g_n'' m_B$ are measures ω' and ω''. ω' is supported on the circle $\{(u, \bar{e}_3), u \perp e_3\}$ and ω'' is supported on the circle $\{(\bar{e}_1, v), e_1 \perp v\}$. It is easy to see that ω', ω'' are the rotation invariant measures on their respective circles. Moreover one can see that all the rotation invariant circular measures are of the form $k\omega'$ or $k\omega''$, $k \in K_3$, where K_3 is the orthogonal group.

Now let φ be a bounded measurable function on B_3. can be shown that the integrals

$$\Phi'(k) = \int \varphi(\xi) \mathrm{d}k\omega'(\xi) \qquad \Phi''(k) = \int \varphi(\xi) \mathrm{d}k\omega''(\xi)$$

may be defined for almost all $k \in K_3$. Moreover these satisfy

$$\int \Phi'(k)\mathrm{d}k = \int \Phi''(k)\mathrm{d}k = \int \varphi(\xi)\mathrm{d}m_B(\xi). \tag{6.3}$$

One way to see this is to regard K_3 as a covering space of B_3.

Recall that K_3 is transitive on $B_3 = G_3/S_3$, so that we can write $B_3 = K_3/W$ for some finite group W. The great circles passing through the identity coset of B_3 correspond to two 1-parameter subgroups, T' and T'', of K_3. It is easy to see that the great circles on B_3 are just the cosets kT' and kT'' taken modulo W. If now φ is an integrable function on B_3, let φ be the corresponding lifted function on K_3. Then

$$\Phi'(k) = \int_{T'} \tilde{\varphi}(kt)\mathrm{d}t, \qquad \Phi''(k) = \int_{T''} \tilde{\varphi}(kt)\mathrm{d}t.$$

and (6.3) follows immediately.

One more fact we shall need is an analogue of the Fatou theorem. In our case this states that for φ a bounded measurable function

$$\int \varphi(k\xi)\mathrm{d}g_n' m_B(\xi) \to \Phi\,(k), \qquad \int \varphi(k\xi)\mathrm{d}g_n'' m_B(\xi) \to \Phi''(k) \qquad (6.4)$$

(at least) in measure, as $n \to \infty$. It follows that if $h(g) = \int \varphi(\xi)\mathrm{d}g m_B(\xi)$ is the corresponding harmonic function, then

$$h(kg_n') \to \Phi'(k), \qquad h(kg_n'') \to \Phi''(k) \qquad (6.5)$$

in measure (of K_3) as $n \to \infty$.

We return now to the harmonic functions satisfying (6.2). If we apply (6.5) combined with (6.2), we see that the 'boundary values' Φ_1', Φ_2' (Φ_1'', Φ_2'') cannot both be positive for a set of k with positive measure. That is

$$\min\,(\Phi_1'(k), \Phi_2'(k)) = 0, \qquad \min\,(\Phi_1''(k), \Phi_2''(k)) = 0 \qquad \text{a.e.}$$

Suppose now that Q_1 and Q_2 almost exhaust B_2 and that ψ_1, ψ_2 are chosen as the characteristic functions of Q_1 and Q_2 respectively. Then $\int[\psi_1(\xi)+\psi_2(\zeta)]\mathrm{d}\theta(\zeta)$ is close to 1. Hence by (6.3), $\int[\Phi_1(k)+\Phi_2(k)]\mathrm{d}k$ is close to 1 as well. Since the two functions in the latter integral are disjoint and bounded by 1, it follows that for most points, if one of the functions vanishes, the other must be close to 1. The same is true for the pair $\Phi_1'(k)$, $\Phi_2''(k)$.

More precisely, we may choose a sequence of pairs Q_1^n, Q_2^n such that the Q_i^n form increasing sequences the sum of whose measures tends to 1. All the functions entering our discussion now depend upon n and they all are non-decreasing as n increases. The limits of the corresponding Φ functions – call them again Φ_i', Φ_i'' – satisfy the following:

$$0 \leqq \Phi_1', \Phi_2', \Phi_1'', \Phi_2'' \leqq 1$$

$$\min(\Phi_1', \Phi_2') = 0 \qquad \min(\Phi_1'', \Phi_2'') = 0$$

$$\int [\Phi_1'(k)+\Phi_2'(k)]\mathrm{d}k = \int [\Phi_1''(k)+\Phi_2''(k)]\mathrm{d}k = 1.$$

These conditions imply that for almost all k, one of $\Phi_1'(k)$, $\Phi_2'(k)$ is 0 and the other is 1. Similarly for $\Phi_1''(k)$, $\Phi_2''(k)$. Now let $\Psi_1(\xi)$, $\Psi_2(\xi)$ be the limits on B_3 of the boundary values of the increasing sequences of harmonic functions. The integral relationships are all maintained because we have bounded convergence. In particular

$$\Phi_i'(k) = \int \Psi_i(\xi)\mathrm{d}k\omega'(\xi), \qquad \Phi_i''(k) = \int \Psi_i(\xi)\mathrm{d}k\omega''(\xi).$$

Now since

$$\Phi_i'(k) = 0 \text{ or } 1, \qquad \Phi_i''(k) = 0 \text{ or } 1$$

we find that outside of an exceptional set of measure 0, Ψ_1 and Ψ_2 must be constant on each great circle. But this cannot be! For if we lift Ψ_1 to the function $\tilde{\Psi}_1$ on K_3, we shall have a function satisfying $\tilde{\Psi}_1(kt) = \tilde{\Psi}_1(k)$ for $t \in T'$ as well as for $t \in T''$. But the set of k for which we have identically $\tilde{\Psi}_1(k'k) = \tilde{\Psi}_1(k')$ form a closed group, and the group generated in K_3 by two distinct one-parameter subgroups is all of K_3. This would force us to the conclusion that Ψ_1 is constant. But it can equal neither 0 nor 1 identically since

$$\int \Psi_1(\xi)\mathrm{d}m_B(\xi) = \lim \int_{Q_1{}^n} \mathrm{d}\theta(\zeta) = \theta(\lim Q_1^n)$$

and lim Q_1^n is an arbitrary measurable subset of B_2. This contradiction shows that (B_2, θ) is not an equivariant image of (B_3, m_B) which completes the proof of Theorem (6.1).

References

[1] Borel, A., Density properties for certain subgroups of semi simple Lie group without compact components, *Annals of Mathematics* 72 (1960) 179–188.

[2] Borel, A., and Harish-Chandra, Arithmetic subgroups of algebraic groups, *Annals of Mathematics* 75 (1962) 485–535.

[3] Chevalley, C., *Theory of Lie Groups*, Princeton, 1946.

[4] Choquet, I. and Deny J., Sur l'équation de convolution $\mu = \mu * \sigma$, *C. R. Acad. Sciences* Paris 250 (1960) 799–801.

[5] Doob, G. Z., *Stochastic Processes*, New York, Wiley, 1953.

[6] Doob, G. Z. Discrete potential theory and boundaries, *Jour. of Math. and Mech.*, 8 (1959) 433–458.

[7] Dudley, R., Random walks on abelian groups, *Proc. Amer. Math. Soc.*, 13 (1962) 447–450.

[8] Dynkin, E. B. and Malyutov M. B., Random walks on groups with a finite number of generators, *Soviet Mathematics (Doklady)*, 2 (1961), 399–402.

[9] Feller, W., *An Introduction to Probability theory and its Applications*, New York, Wiley, 1950.

[10] Feller, W., Boundaries induced by non-negative matrices, *Trans. Amer. Math. Soc.* 83 (1956), 19–54.

[11] Furstenberg, H., A Poisson formula for semi-simple Lie groups, *Annals of Mathematics* 77 (1963) 335–386.

[12] Furstenberg, H., Noncommuting random products, *Trans. Amer. Math. Soc.*, 108 (1963), 377–428.

[13] Greenander, U., *Probabilities on Algebraic Structures*, Uppsala, Wiley 1963.

[14] Hunt, I., Semigroups of measures on a Lie group, *Trans. Amer. Math. Soc.*, 81 (1956), 264–293.

[15] Helgason, S., *Differential Isometry and Symmetric Spaces*, New York, Academic Press, 1962.

[16] Kesten, H., Symmetric random walks on groups, *Trans. Amer. Math. Soc.*, 92 (1959) 336–354.

[17] Loeve, M., *Probability Theory*, New York, Van Nostrand 1964.

[18] Malcev, A. I., On a class of homogeneous spaces, *Amer. Math. Soc. Translation* 39 (1949).

[19] Moore, C. C., Ergodicity of flour on homogeneous spaces, *Amer. Jour. of Math.* 88 (1966) 154–178.

[20] Mostow, I. D., Factor spaces of solvable groups, *Annals of Mathematics*, 60 (1954), 1–27.

[21] Newman, M. The structure of some subgroups of the modular group. *All. Jour. of Math.* 6 (1962) 480–487.

[22] Spitzer, F., *Principles of Random Walk*, London, Van Nostrand 1964.

The technique of using random measures and random sets in harmonic analysis

by J.-P. Kahane

UNIVERSITÉ DE PARIS
INSTITUT HENRI POINCARÉ
PARIS, FRANCE

1. Introduction

From the point of view of classical analysis, a nowhere differentiable function is a rather strange object. But the best known random function – the brownian motion function – is a. s. nowhere differentiable.

Such an opposition occurs in many problems in analysis. The problem is to find a function, or a measure, or a distribution, or a set, enjoying given properties. Sometimes it is hard or even impossible to give a solution by means of an explicit construction, but it is easy to define a random function (or measure, or distribution, or set) such that almost surely the given properties are satisfied. Our purpose is to give a series of examples of this situation.

The problems of analysis we shall consider are the following:

- divergence of Fourier series (Section 2)
- classes of conjugate functions, and radial behaviour (Section 3)
- perfect sets of multiplicity for the trigonometric expansion (Sections 5, 6, 7)
- Fourier transforms of singular measures (Sections 5, 6, 7)
- different dimensions of perfect sets, and arithmetical properties (Sections 7, 8)
- Hölder conditions and Peano curves (Section 9)
- range of a function restricted to a perfect set (Sections 9, 10)
- sets of non spectral synthesis (Section 11)

Actually we are not so much interested in these problems than in the tools we shall use. These tools are very simple and natural random objects, such as a sum of randomly distributed point masses on the circle (Sections 1, 2), or the image of the Lebesgue measure by a positive stochastic process with independent increments (Section 7), or the image of a closed set by a gaussian stochastic process (Section 9), or the set of zeros of a gaussian stochastic process (Section 10). These random objects are very interesting by themselves, and deserve attention even without further application; we did not escape the temptation to develop an example of a random closed set on the circle which is completely useless until now (Section 4).

The technique of using a probabilistic argument instead of an explicit construction was introduced by N. Wiener about 1930 – as far as we know –. The question was to give an example of a bounded function f on the line such that

$$\varphi(x) = \lim_{T\to\infty} \frac{1}{2T}\int_{-T}^{T} f(t+x)\bar{f}(t)\mathrm{d}t$$

exists for each x and φ is the Fourier transform of a continuous function on the line. N. Wiener gave an explicit example, and proves also that if we define $f(t) = a_n$ when $n \leqq t < n+1$ $(n = \ldots -1, 0, 1, \ldots)$ and choose $a_n = +1$ or -1 at random (with probability $\frac{1}{2}$, each choice being independent from the other), a.s. $\varphi(x) = \sup(0, 1-|x|)$ [22].

The method was developed about the same time by Paley and Zygmund. A spectacular result of their is the following: if $\sum_1^\infty |c_n|^2 = \infty$, almost all series $\sum_1^\infty \pm c_n \cos nt$ are not Fourier-Lebesgue series. In this case, no explicit construction is known [17]. Another result of Paley and Zygmund is given in Section 3.

Later on, Salem used this idea in order to investigate sets of multiplicity and singular measures. His method and results are given in Section 6. Here again, some important theorems can be proved only by means of probabilistic argument.

Many other examples will be given in [13].

The results which are underlined in Sections 4, 8, 9, 10, are published here for the first time.

2. Sums of Independent Point-Masses on the Circle

Let ρ_j $(j = 1, 2, \ldots)$ be a sequence of positive numbers, such that

$$\sum_1^\infty \rho_j < \infty. \tag{2.1}$$

We define the circle to be the set of real numbers defined modulo 1, and we consider a sequence of independent random variables ω_j equidistributed on the circle. By δ_{ω_j} we denote the unit mass at ω_j, and we consider the random measure

$$\sum_1^\infty \rho_j \delta_{\omega_j}. \tag{2.2}$$

We may use (2.2) in order to prove the existence of a Fourier-Lebesgue series that diverges almost everywhere. The first proof is due to Kolmogorov (1923) and consists in giving an explicit example [23, p. 305].

Let us introduce a few notations. As usual

$$D_m(t) = \sum_{-m}^{m} \mathrm{e}^{2\pi i n t} = \frac{\sin(2m+1)\pi t}{\sin \pi t}$$

is the Dirichlet kernel. A Fourier-Lebesgue series is a series

$$\sum_{-\infty}^{\infty} a_n \mathrm{e}^{2\pi i n t}, \quad \text{where} \quad a_n = \int f(t)\mathrm{e}^{-2\pi i n t}\,\mathrm{d}t \tag{2.3}$$

for some $f \in L^1$, the integration being taken over the circle. A Fourier-Stieltjes series is a series

$$\sum_{-\infty}^{\infty} c_n \mathrm{e}^{2\pi i n t}, \quad \text{where} \quad c_n = \int \mathrm{e}^{-2\pi i n t}\,\mathrm{d}\mu(t) \tag{2.4}$$

for some measure $\mathrm{d}\mu$. The 'partial sums' of (2.3) and (2.4) are

$$S_m(f, t) = \sum_{-m}^{m} a_n \mathrm{e}^{2\pi i n t} = \int D_m(t-s) f(s)\,\mathrm{d}s$$

$$S_m(d\mu, t) = \sum_{-m}^{m} c_n \mathrm{e}^{2\pi i n t} = \int D_m(t-s)\,\mathrm{d}\mu(s).$$

We say that (2.4) is strongly divergent at t if

$$\overline{\lim_{m\to\infty}}\, S_m(d\mu, t) = \infty. \tag{2.5}$$

The problem of finding an a.e. divergent Fourier-Lebesgue series can be reduced to the problem of finding a positive measure with an a.e. strongly divergent Fourier-Stieltjes series. For, if (2.5) holds a.e., there exists a positive sequence λ_n tending to zero such that

$$\overline{\lim_{m\to\infty}}\, \lambda_m S_m(\mathrm{d}\mu, t) = \infty \text{ almost everywhere.}$$

Now we use the fact that, given a positive convex sequence γ_n tending to zero,

$$\varphi(t) = \gamma_0 + \sum_0^\infty 2\gamma_n \cos 2\pi nt$$

is a positive summable function, and

$$S_m(\varphi, t) - \gamma_m D_m(t)$$

is positive too [23, p. 183]. We define the γ_n so that $\gamma_n \geqq \lambda_n$ and we consider

$$f(t) = \int \varphi(t-s) \mathrm{d}\mu(s).$$

That is a summable function, and we have

$$\begin{aligned} S_m(f, t) &= \int S_m(\varphi, t-s) \mathrm{d}\mu(s) \\ &\geqq \int \gamma_m D_m(t-s) \mathrm{d}\mu(s) \\ &\geqq \lambda_m S_m(\mathrm{d}\mu, t) \end{aligned}$$

if the last term is positive. Therefore

$$\overline{\lim_{m\to\infty}} S_m(f, t) = \infty \text{ almost everywhere.}$$

And now the random measure (2.2) comes in season. For we have the following simple theorem: *if*

$$\sum_1^\infty \rho_j \log \frac{1}{\rho_j} = \infty, \tag{2.6}$$

a. s. the Fourier-Stieltjes series of (2.2) *is strongly divergent almost everywhere.*

The proof is as follows. We write $\mathrm{d}\mu$ for the sum of (2.2). Since all translates of $\mathrm{d}\mu$ have the same distribution, the probability of (2.5) does not depend on t. Therefore, the question is to prove that a. s.

$$\overline{\lim_{m\to\infty}} S_m(\mathrm{d}\mu, 0) = \infty. \tag{2.7}$$

We have

$$S_m(\mathrm{d}\mu, 0) = \sum_1^\infty \rho_j D_m(\omega_j)$$

and choosing $|\omega_j| \leqq \frac{1}{2}$, as we can,

$$D_m(\omega_j) = \frac{\sin 2\pi m\omega_j}{\pi\omega_j} + 0(1).$$

If we define

$$S_m^* = \sum_1^\infty \rho_j \frac{\sin 2\pi\omega_j m}{\pi\omega_j},$$

(2.7) is the same as $\overline{\lim}_{m\to\infty} S_m^* = \infty$. Now, given any number of integers $n_0, n_1, n_2, \ldots, n_k$ ($n_j \geqq 0$ or $\leqq 0$ and $|n_1|+|n_2|+\ldots+|n_k| \neq 0$), we have a. s.

$$n_0 + n_1\omega_1 + n_2\omega_2 + \ldots + n_k\omega_k \neq 0.$$

It follows that a.s. $1, \omega_1, \omega_2, \ldots, \omega_j, \ldots$ are linearly independent. According to the Kronecker theorem,

$$\text{a.s.} \sup_m \sum_1^v \rho_j \frac{\sin 2\pi\omega_j m}{\pi\omega_j} = \sum_1^v \frac{\rho_j}{\pi|\omega_j|}$$

It follows easily that for each v (2.7) holds a.s. if (and only if)

$$\text{a.s.} \sum_1^\infty \frac{\rho_j}{|\omega_j|} = \infty. \tag{2.8}$$

A necessary and sufficient condition for (2.8) is

$$\sum_1^\infty \mathscr{E}\left(\min\left(\frac{\rho_j}{|\omega_j|}, 1\right)\right) = \infty$$

and that is nothing but (2.6). Therefore (2.6) implies that the Fourier-Stieltjes series of (2.2) is strongly divergent almost everywhere, q.e.d.

We only mention the strongest result that can be obtained in this way: *given an increasing sequence of positive integers* $m_1 < m_2 < \ldots$ *and a positive function* $\psi(x) = o(\log\log x)$ $(x \to \infty)$, *there exists a summable function* f *on the circle, such that*

$$\overline{\lim_{k\to\infty}} \frac{1}{\psi(k)} S_{m_k}(f, t) = \infty$$

almost everywhere [11], [13].

Let us add two remarks on the divergence of Fourier series.

1. Kolmogorov constructed also (1926) a Fourier-Lebesgue series that diverges everywhere [23, p. 310]. That suggests the following question: how is it possible to choose the ρ_j in order that the Fourier-Stieltjes series (2.2) is a.s. strongly divergent everywhere? A modification of the proof shows that again (2.6) is a necessary and sufficient condition (oral communication of Y. Katznelson).

2. Carleson proved recently that a Fourier series of a square summable function f converges almost everywhere to f. If we suppose $f(\log|f|)^{1+\delta} \in L^1$ instead of $f \in L^2$ $(\delta > 0)$, the conclusion is $S_n(f, t) = o(\log\log n)$ a. e. It may be conjectured that the same holds when $f \in L^1$; in this case it would be the best possible result.

3. Sums of Independent Signed Point-Masses on the Circle

Again we consider a sequence of positive numbers ρ_j $(j = 1, 2, \ldots)$. Here we assume

$$\sum_1^\infty \rho_j^2 < \infty \tag{3.1}$$

and we consider the random series

$$\sum_{j=1}^\infty \varepsilon_j \rho_j \delta_{\omega_j} \tag{3.2}$$

where the ε_j and the ω_j are independent random variables, $\varepsilon_j = +1$ or -1 with the same probability, and ω_j equidistributed on the circle. We can define the Fourier coefficients of (3.2) is a formal way:

$$c_n = \sum_{j=1}^\infty \varepsilon_j \rho_j e^{-2\pi i n \omega_j}.$$

Each c_n is a symmetric complex random variable, and

$$\mathscr{E}(|c_n|^2) = \sum_{j=1}^\infty \rho_j^2.$$

Therefore $p(|c_n| > |n|) = 0(|n|^{-2})$. Therefore

$$\text{a.s.} \quad c_n = 0(|n|).$$

This implies that a.s. the series

$$\sum_{-\infty}^{\infty} c_n e^{2\pi i n t}$$

is the Fourier-Schwartz series of a distribution on the circle, which we denote by T.

Now let us suppose $\sum_1^\infty \rho_j^2 = 1$, and let us consider an infinitely differentiable function f on the circle

$$f = \sum_{-\infty}^{\infty} a_n e^{2\pi int}.$$

Then

$$\langle T, f\rangle = \sum_{-\infty}^{\infty} c_{-n} a_n = \sum_{j=1}^{\infty} \varepsilon_j \rho_j f(\omega_j)$$

is a symmetric random variable, and

$$\mathscr{E}(|\langle T, f\rangle|^2) = \int |f(t)|^2 \mathrm{d}t.$$

The mapping $T: f \to \langle T, f\rangle$ can be extended as a linear norm-preserving mapping from L^2 (circle) into L^2 (probability space). One may think of T as an analogue of the homogeneous chaos of N. Wiener. Actually, when all the ρ_j are very small, $\langle T, f\rangle$ is almost a gaussian variable, and therefore T is a good approximation for a random distribution which can be extended as a linear norm-preserving mapping from L^2 (circle) into a Hilbert space of symmetric gaussian variables; such a random distribution is precisely the homogeneous chaos on the circle, and it gives by integration the Wiener function of the brownian motion.

The random distributions (3.2) prove to be useful in a question about conjugate harmonic functions in the unit disc. Given an increasing function $\chi(x)$ on $[0, \infty]$, such that $\lim_{x\to 0} \chi(0) = 0$ and $\lim_{x\to\infty} \chi(x) = \infty$, we denote by h_χ the class of all harmonic functions u in the unit disc $|z| \leqq 1$ such that the mean values of $\chi(|u|)$ on circles $|z| = r$ are uniformly bounded with respect to r $(0 < r < 1)$. Writing $u(z)$ instead of $u(\operatorname{Re} z, \operatorname{Im} z)$, this condition is

$$\int \chi(|u(re^{2\pi it})|)\mathrm{d}t = 0(r) \qquad (r \to 1).$$

We denote by $\tilde{u}$ the conjugate function of u, and we call χ a scale function. The question is as follows: *given a scale function χ, define a scale function ψ such that $u \in h_\chi$ implies $\tilde{u} \in h_\psi$?*

The main positive results are the following: 1) if $\chi(x) = x^p$,

$p > 1$, one may choose $\psi(x) = x^p$; in other words, the class h_χ is stable under conjugation (Marcel Riesz) 2) if $\chi(x) = x$, one may choose $\psi(x) = x^p$ with $p < 1$ (Kolmogorov) [23, p. 253].

The negative results express that Kolmogorov's theorem is best possible in several directions: 1) if $\chi(x) = x$, one cannot choose $\psi(x) = x$ (in other words, h_ψ is not stable under conjugation) 2) *if* $\chi(x) = o(x)$ $(x \to \infty)$, *there does not exist any scale function* ψ. The first result is easy .The second was obtained by Paley and Zygmund in 1935 [18]. Beforehand, Littlewood had proved that if $\chi(x) = x^p$, $p < 1$, one cannot choose $\psi(x) = x^q$ for any $q > 0$.

The method of Littlewood is to define u and $\tilde{u}$ as the real and imaginary parts of

$$\sum_1^\infty \frac{z^{\lambda_n}}{1-z^{2\lambda_n}},$$

where λ_n is a rapidly increasing sequence of integers. The method of Paley and Zygmund consists in randomizing Littlewood's example; they define u and $\tilde{u}$ as the real and imaginary parts of

$$\sum_1^\infty \varepsilon_n \frac{z^{\lambda_n}}{1-z^{2\lambda_n}},$$

where the $\varepsilon_n = \pm 1$ are independent random variables as in (3.2). It is slightly easier, and more natural, to prove their theorem on randomizing one step more, i.e. to define u and $\tilde{u}$ as the random harmonic functions which are obtained by convolving (3.2) with the Poisson kernel and the conjugate Poisson kernel. We only sketch the proof; the details will be found in [13].

We write

$$P_r(t) = \sum_{-\infty}^{\infty} r^{|n|} e^{2\pi i n t} = \frac{1-r^2}{1-2r\cos 2\pi t + r^2}$$

$$\tilde{P}_r(t) = 2\sum_1^\infty r^n \sin 2\pi n t = \frac{2r\sin 2\pi t}{1-2r\cos 2\pi t + r^2}$$

$$u(re^{2\pi i t}) = \sum_1^\infty \varepsilon_j \rho_j P_r(t-\omega_j)$$

$$\tilde{u}(re^{2\pi i t}) = \sum_1^\infty \varepsilon_j \rho_j \tilde{P}_r(t-\omega_j).$$

It is easy to check that u and $\tilde{u}$ are a.s. defined in the unit disc $|z| < 1$, harmonic and conjugate. Suppose we are given two scale functions $\chi(x) = o(x)$ and $\psi(x)$.

The first step is to choose to ρ_j so that

$$\sum_1^\infty \rho_j = \infty \tag{3.3}$$

and $u \in h_\chi$. That is not difficult because

$$\int \chi(P_r(t))\mathrm{d}t = o(1) \qquad (r \to 1).$$

The second step consists in proving that, given any $A > 0$ (arbitrarily large), there exists $r < 1$ such that the event

$$\text{meas. } \{t \| \tilde{u}(re^{2\pi it})| > A\} > \tfrac{1}{100} \tag{3.4}$$

has probability $> \frac{1}{100}$. That implies readily a.s. $\tilde{u} \notin h_\psi$. In order to look at the event (3.4), it is convenient to write $\varepsilon = (\varepsilon_1, \varepsilon_2, \ldots, \varepsilon_n, \ldots)$, $\omega = (\omega_1, \omega_2, \ldots, \omega_n, \ldots)$, $\tilde{u}(z) = \tilde{u}(z, \varepsilon, \omega)$. For each given r, t, ω, the ε-set where

$$|\sum_1^\infty \varepsilon_j \rho_j \tilde{P}_r(t-\omega_j)| > \tfrac{1}{2}(\sum_1^\infty \rho_j^2 \tilde{P}_r^2(t-\omega_j))^{\frac{1}{2}}$$

has measure $> \frac{1}{10}$. For each given r, t, the ω-set where

$$\sum_1^\infty \rho_j^2 \tilde{P}_r^2(t-\omega_j) > 4A^2$$

has a measure independent of t, and larger than $\frac{1}{2}$ if r is large enough (because of (3.3)), $r = r_A$, say. Therefore, the (ε, ω, t)-set where

$$|\tilde{u}(r_A e^{2\pi it})| > A$$

has measure $> \frac{1}{20}$. On the other hand, denoting by α the probability of (3.4), this set has measure $\leqq \alpha + \frac{1}{100}(1-\alpha)$. Therefore $\alpha > \frac{1}{100}$, q.e.d.

Actually, the proof gives a little more than needed, namely that, given the ρ_j such that (3.3) holds, there exists a sequence r_n tending to 1 such that the event

$$\text{meas. } \{t \| u(r_n e^{2\pi it})| > n\} > \tfrac{1}{100} \text{ for infinitely many } n \tag{3.5}$$

has probability $> \frac{1}{100}$. Therefore, *given a scale function* $\chi(x) = o(x)$, *there exists* $u \in h_\chi$ *such that* $\tilde{u} \notin h_\psi$ *for any scale function* ψ.

4. Uncovered Sets on the Circle

Suppose we are given a sequence of positive numbers l_j $(0 < l_j < 1; j = 1, 2, \ldots)$. Each l_j is considered as the length of an open interval I_j on the circle. The union

$$E = \bigcup_1^\infty I_j \tag{4.1}$$

is an open set. We denote by F the complementary set. What can we say about F?

Obviously, this question has no interesting answer. Let us randomize the problem. Given the l_j, we consider the I_j as randomly distributed on the circle; the length of I_j is l_j, the center is ω_j, and $\omega_1, \omega_2, \ldots$ is a sequence of independent random variables equidistributed on the circle as in Section 2 and Section 3. Now E and F are random sets on the circle. It is easy to see that F is a.s. nowhere dense. We may ask conditions on the l_j such that a.s. F has measure zero, or F is empty, or F has a Hausdorff dimension less than α. Now the question is meaningful, and actually we shall give some partial answers.

Given t on the circle, the events $t \in I_j$ are independent, and

$$p(t \in I_j) = l_j.$$

As an application of the Borel-Cantelli lemma,

$$\sum_1^\infty l_j = \infty \Leftrightarrow \text{a.s. meas. } F = 0.$$

It is not so easy to give necessary and sufficient conditions for F to be empty. The question was asked by Dvoretsky. Typical answers are the following: if $l_j = 1+\varepsilon/j$ $(\varepsilon > 0)$, a.s. $F = \emptyset$[7]; if $l_j = 1-\varepsilon/j$ $(\varepsilon < 0)$, $p(F \neq \emptyset) > 0$[3]. For $l_j = 1/j$, the answer is not known, but Billard succeeded in proving that F is a.s. at most countable [3].

Surprisingly, the question on Hausdorff dimension has a rather simple answer. *Let us assume* $l_1 \geqq l_2 \geqq l_3 \geqq \ldots$, *and define*

$$\kappa = \overline{\lim_{n\to\infty}} \left(\frac{1}{\log n} \sum_1^n l_j\right). \tag{4.2}$$

If $\kappa > 1$, *a.s.* $F = \emptyset$. *If* $\kappa \leqq 1$, *a.s.* $\dim F \leqq 1-\kappa$ *and* $p(\dim F \geqq 1-\kappa-\varepsilon) > 0$ *for each* $\varepsilon > 0$.

We shall prove this theorem by using a modification of the method of Billard. We need a series of steps. To avoid trivialities, we suppose $\kappa > 0$.

1. Let I be an interval of length $|I|$ on the circle, and μ an integer such that $l_\mu < |I|$ and $l_\mu < \frac{1}{2}$. For each $\nu > \mu$, we have

$$p(I \not\subset \bigcup_{j=\mu+1}^{\nu} I_j) \leqq (1+4(\nu-\mu)|I|) \exp\,(-\sum_{\mu+1}^{\nu} l_j).$$

For, if I is not contained in the union of the I_j ($j = \mu+1, \ldots \nu$), either no I_j contains the left extremity of I or there exists an I_j which intersects I and such that the right extremity of I_j does not belong to any I_k ($k \neq j$). Therefore

$$p(I \not\subset \bigcup_{\mu+1}^{\nu} I_j) \leqq \prod_{\mu+1}^{\nu}(1-l_j)+\sum_{j=\mu+1}^{\nu}\frac{l_j+|I|}{1-l_j}\prod_{k=\mu+1}^{\nu}(1-l_k),$$

which implies (4.3) readily.

2. As a consequence, we have the first part of the theorem: if $\kappa > 1$, a. s. $F = \emptyset$.

3. In the particular case $l_j = \alpha/j$, we may choose as ν the largest integer such that $(\nu-\mu)|I| \leqq 1$, and we obtain

$$p(I \not\subset \bigcup_{\mu+1}^{\infty} I_j) \leqq 5(\mu+1)^\alpha|I|^\alpha. \tag{4.4}$$

4. Suppose again $l_j = \alpha/j$, and let A be a set of Hausdorff measure 0 in the dimension α (i.e. A can be covered by a countable union of intervals I such that $\sum |I|^\alpha$ is arbitrarily small). Then a.s. $A \subset E$.

For, according to (4.4), $p(A \not\subset \bigcup_{\mu+1}^{\infty} I_j) = 0$.

5. Suppose now that A is a closed set on the circle and $d\sigma(t)$ a positive measure carried by A, of total mass 1. We define

$$\alpha_n = \iint_{|u-v|\leqq l_n} d\sigma(u)d\sigma(v).$$

If we have

$$\sum_{n=1}^{\infty} \alpha_n l_n \exp\,(\sum_{j=1}^{n} l_j) < \infty, \tag{4.5}$$

then $p(A \not\subset E) > 0$.

We only sketch the proof, which is given in [3, *p.* ?] in the case A is the whole circle and $d\sigma(t)$ is the Lebesgue measure. Let us denote by $\chi_j(t-\omega_j)$ the characteristic function of I_j (= 1 on I_j and 0 outside). Then

$$\prod_1^\nu (1-\chi_j(t-\omega_j))$$

is the characteristic function of $F_\nu = \complement \bigcup_1^\nu I_j$. An easy computation gives

$$\mathscr{E}(\sigma(F_\nu)) = \prod_1^\nu (1-l_j)$$
$$\mathscr{E}(\sigma^2(F_\nu)) = \int \prod_1^\nu (1-2l_j+\xi_j(t))d\sigma(t),$$

where $\xi_j = \chi_j * \chi_j$ (convolution of χ_j by itself). We may write

$$\mathscr{E}(\sigma^2(F_\nu)) = \gamma_\nu \prod_1^\nu (1-l_j)^2$$

and (4.5) implies that γ_ν is uniformly bounded, say $\gamma_\nu \leqq \gamma$. Since

$$\mathscr{E}^2(\sigma(F_\nu)) \leqq p(\sigma(F_\nu) > 0)\mathscr{E}(\sigma^2(F_\nu))$$

we have

$$p(F_\nu \cap A \neq \emptyset) \geqq p(\sigma(F_\nu) > 0) \geqq \gamma^{-1}.$$

Therefore

$$p(F \cap A \neq \emptyset) = p(A \not\subset E) > 0.$$

6. As a particular case, suppose that A has a positive measure in the Hausdorff dimension β. According to a theorem of Frostman (cf. for example [15, p. 27]), A carries a measure $d\sigma$ such that $\sigma(I) \leqq C|I|^\beta$ for each interval I, C being independent of I. Therefore the condition

$$\sum_{n=1}^\infty l_n^{1+\beta} \exp\left(\sum_1^n l_j\right) < \infty \tag{4.6}$$

implies $p(A \not\subset E) > 0$.

7. Let us consider κ, defined in (4.2), and again a closed set A. If $\kappa < \dim A$, we have $p(A \not\subset E) > 0$.

For $\kappa < \infty$ implies $l_n = 0(\log n/n)$ and $\exp\left(\sum_1^n l_j\right) = 0(n^{\kappa+\varepsilon})$ for each $\varepsilon > 0$. Therefore $\kappa < \beta$ implies (4.6).

8. In particular $\kappa < 1$ implies $p(F \neq \emptyset) > 0$.

9. In the particular case $l_j = \alpha/j$ and $\dim A > \alpha, p(A \not\subset E) > 0$.

10. Let us give another formula for κ. Given a countable family of points in $[1, \infty[$, Λ, we define $n(r)$ as the number of points of Λ (each of them counted with its order of multiplicity) in $[1, r[$, and

$$d = \overline{\lim_{r\to\infty}} \left(\frac{1}{\log r}\int_1^r \frac{dn(t)}{t}\right);$$

d is called the upper logarithmic density of Λ. Then κ is the upper logarithmic density of $\{1/l_1, 1/l_2, \ldots\}$.

For writing $\Lambda = \{1/l_1, 1/l_2, \ldots\}$, we have

$$d = \overline{\lim_{n\to\infty}} \left(\frac{1}{\log(1/j_n)}\sum_1^n l_j\right).$$

Since $l_n = 0(\log n/n)$, we have $d \leqq \kappa$. On the other hand, given $\varepsilon > 0$, we have

$$\sum_1^n \left(l_j - \frac{\kappa-\varepsilon}{j}\right) > 0, \qquad l_n > \frac{\kappa-\varepsilon}{n}$$

for infinitely many n; therefore $d \geqq \kappa - \varepsilon$. Therefore $d = \kappa$.

11. From now on, given a countable family L on $[0, 1[$, we define $\kappa(L)$ as the upper logarithmic density of $\{1/l\}_{l\in L}$ (that does not depend on any ordering of L). We write L_0 for the given family $\{l_n\}$, L_α for the family $\{\alpha/n\}$ $(n = 1, 2, \ldots)$, and $L_0 \cup L_\alpha$ for their union (the order of multiplicity of each point being the sum of its orders in L_0 and in L_α). Then

$$\kappa(L_0) = \kappa, \qquad \kappa(L_0 \cup L_\alpha) = \kappa + \alpha.$$

That follows immediately from the preceding step.

12. Given L_0 as before, and K a compact set on the circle, we say that L_0 is covering for K if a.s. $E = \bigcup_1^\infty I_j$ contains K. We have a similar definition for any family L instead of L_0.

According to the steps 2 and 11, $L_0 \cup L_\alpha$ is covering for the circle if $\kappa + \alpha > 1$. Therefore, for almost all choices of the I_j, L_α is covering for F. On the other hand, we know (step 9) that L_α is not covering for any closet set whose Hausdorff dimension exceeds α. Therefore a. s. $\dim F \leqq \alpha$. Therefore a. s. $\dim F \leqq 1 - \kappa$.

13. According to the steps 8 and 11, $L_0 \cup L_\alpha$ is not covering

for the circle if $\kappa+\alpha < 1$. Therefore, for many choices of the I_j (we mean: with a positive probability), L_α is not covering for F. On the other hand (step 4), L_α is covering for any closed set whose Hausdorff dimension is less than α. Therefore $p(\dim F > \alpha) > 0$. That ends the proof of the theorem.

A more elegant formulation can be obtained if we consider

$$E_\infty = \overline{\lim_{j\to\infty}} I_j = \bigcap_{\mu=1}^{\infty} \bigcup_{j=\mu+1}^{\infty} I_j$$

instead of E, and $F_\infty = \complement E_\infty$ instead of F. F_∞ is the set of points which are covered by finitely many I_j. A. s. F_∞ is a countable union of nowhere dense closed sets, i.e. a set of Baire's first category [2]. Applying the zero-one law, we obtain: *if* $\kappa > 1$, *a.s.* $F_\infty = \emptyset$; *if* $\kappa \leqq 1$, *a. s.* $\dim F_\infty = 1-\kappa$.

5. Problems on Perfect Sets and Singular Measures

Several important problems in harmonic analysis are of the following type: construct a measure, or a distribution, carried by a 'thin' set, with a Fourier transform 'small' at infinity.

A classical example of this situation is the problem of uniqueness and multiplicity for trigonometric series: given a closed set E on the circle, does there exist a trigonometric series that converges to zero out of E and that is not identically zero? If the answer is positive, E is a set of multiplicity, or M-set. If the answer is negative, E is a set of uniqueness, or U-set. The empty set is a U-set (G. Cantor) and each set of positive measure is a M-set. The classification of closed sets of measure zero into U-sets and M-sets has been studied by Rajchman, Menšov, Nina Bary, Salem, Pyatetski-Shapiro, Zygmund ([23, chap. IX and XII], [15, chap. V and VI]). A reformulation of the problem is the following: given E, does there exist a Schwartz distribution, carried by E, $\not\equiv 0$, whose Fourier transform tends to zero at infinity? If E carries a measure $\not\equiv 0$ whose Fourier transform tends to zero at infinity, we say that E is a set of multiplicity in the restricted sense, or M_0-set. The first example of a non trivial M_0-set was given by Menšov in 1916.

Another classical problem is the following: given $\alpha > 0$, does there exist a purely singular measure whose Fourier transform is

$0(|u|^{-\alpha})$ $(|u| \to \infty)$? Obviously from the Plancherel theorem, the answer is negative if $\alpha > \frac{1}{2}$. Littlewood proved in 1936 that the answer is positive when α is small enough, and in 1938, Wiener and Wintner succeeded in proving the same for any $\alpha < \frac{1}{2}$. The best result in this direction is due to Ivašev-Musatov [6]: if $\varphi(u)$ is a positive function defined on $[1, \infty[$, such that $u^\alpha \varphi(u)$ is monotonic for any $\alpha > 0$, and $\int_0^\infty \varphi^2(u)\mathrm{d}u = \infty$, there exists a measure $\mathrm{d}\mu$, carried by a closed set of Lebesgue measure zero, whose Fourier transform satisfies $\hat{\mu}(u) = 0(\varphi(|u|))$ $(|u| \to \infty)$.

In the following sections we shall study a more comprehensive question. Given a positive function φ defined on $[1, \infty[$, let $M_{(\varphi)}$ be the class of measures $\mathrm{d}\mu$ on the line such that $\hat{\mu}(u) = o(\varphi(|u|))$ $(|u| \to \infty)$; if $\varphi(u) = u^\alpha$, we write M_α instead of $M_{(\varphi)}$. A compact set E on the line will be called a $M_{(\varphi)}$-set (or M_α-set) if it carries a measure $\mathrm{d}\mu \in M_{(\varphi)}$ (or M_α). The question is to obtain as many examples of $M_{(\varphi)}$-sets as possible. In some cases we prescribe a geometrical property, for example to be a symmetric set of a Cantor type; or a dimension property, for example to have a given Hausdorff dimension; or an arithmetical property, for example to be independent over the rationals.

The first systematic study in this direction is due to Salem. We shall indicate his method in the following section. The main point is that no explicit example is given, but a random set is constructed and is proved to be a. s. a $M_{(\varphi)}$-set ([19], [20], see also [15, chap. VIII]). The main results of his are the following: 1. 'Almost all' symmetric perfect sets are M_0-sets [19]. 2. In another sense, 'almost all' symmetric perfect sets are M_α-sets for some $\alpha > 0$. The precise statement will be given later [20]. 3. From the definition of the capacitarian dimension (which is the same as the Hausdorff dimension) follows immediately that a M_α-set has dimension $\geqq 2\alpha$. Conversely, for each $\alpha < \frac{1}{2}$, there exist M_α-sets whose dimension is exactly 2α [20].

The first result had important applications. It gives examples of M_0-sets which do not generate the additive real group. By a modification of the construction, Rudin gave examples of M_0-sets which are independent over the rationals [15, p. 103], and Varopoulos extended the construction to general locally compact non discrete abelian groups [21]. That turns out to be a very good tool for the investigation of the convolution measure algebra on the line and other groups.

As an application of the second result, we just indicate the following: given a positive sequence λ_n such that $(\lambda_{n+1}/\lambda_n) > q > 1$ $(n = 1, 2, \ldots)$, 'almost all' symmetric perfect sets contain some u, for which the sequence $\lambda_n u$ has no distribution modulo 1 [5].

Applications of the third result are given in [15, p. 112] and [9, p. 134]. This result can be restated in the following way. Let us define the Fourier dimension of a compact set E as the supremum of the β such that E carries a measure $\not\equiv 0$ $d\mu \in M_{\beta/2}$; we denote it by $\mathscr{F}$-dim E. If $\dim E < 1$, $\mathscr{F}\text{-dim } E \leqq \dim E$. Conversely, given any $\beta \in]0, 1[$, there exist E such that $\mathscr{F}\text{-dim } E = \dim E$. Here Salem's method is rather complicated, and we shall see later a very natural argument in order to obtain as well this result and Rudin's result, using positive stochastic processes with independent increments.

6. The Salem Method

Since the first and third result above are explained in [15], we shall explain the method in the second case.

Given a sequence ξ_k, $0 < \xi_k < \frac{1}{2}$ $(k = 1, 2, \ldots)$, we define $E = E(\{\xi_k\})$ in the following way. From a given interval I we remove the interval with same center and length $(1-2\xi_1)|I|$; we obtain two intervals of length $\xi_1|I|$. With each of those intervals we repeat the dissection, by using ξ_2 instead of ξ_1; we obtain four intervals of length $\xi_1\xi_2|I|$. At the nth step, repeating the procedure, we have a set E_n which consist in 2^n intervals of length $\xi_1\xi_2\ldots\xi_n$. The set E is defined as $\bigcap_{n=1}^{\infty} E_n$.

Now we randomize the definition, by choosing ξ_k at random between a_k and b_k, where a_k and b_k are two given sequences; we suppose that the random variables ξ_k are independent, and each ξ_k is equidistributed on $[a_k, b_k]$. Moreover, we suppose that a_k and b_k are not too close, namely

$$b_k - a_k \geqq \frac{1}{\omega(k)}, \qquad \omega(k)\to, \qquad \log \omega(k) = o(k). \tag{6.1}$$

Here is Salem's theorem: *if*

$$\underline{\lim}\,(a_1 \ldots a_k)^{1/k} = \delta > 0, \tag{6.2}$$

a. s. E is a M_α-set for some $\alpha = \alpha(\delta) > 0$.

In order to prove the theorem, we may suppose E constructed on $[-l, l]$ with $l < \pi$. E carries a measure $d\mu$ whose Fourier transform is

$$\hat{\mu}(u) = \prod_{k=1}^{\infty} \cos(ln\, \xi_1 \xi_2 \ldots \xi_{k-1}(1-\xi_k)).$$

The Fourier coefficients of $d\mu$ are $\hat{\mu}(n)$, and because $l < \pi$

$$\hat{\mu}(n) = o(|n|^{-\alpha}) \Leftrightarrow \hat{\mu}(u) = o(|u|^{-\alpha}). \tag{6.3}$$

Now the idea is to define two positive numbers, s (large) and β (small) such that

$$\mathscr{E}\Big(\sum_{n=-\infty}^{\infty} |\hat{\mu}(n)|^s |n|^\beta\Big) < \infty. \tag{6.4}$$

Choosing $\alpha = \beta/s$, (6.3) and (6.4) imply that a.s. E is a M_α-set.

The computation of (6.4) is not very difficult. We define

$$p = p(n) = \left[\frac{\log n}{2 \log(1/\delta)}\right] \qquad ([\ ] = \text{integral part}) \tag{6.5}$$

and, for a given n, we use the estimate

$$\mathscr{E}(|\hat{\mu}(n)|^s) \leqq \mathscr{E}\Big(\prod_{k=1}^{p} |\cos ln\, \xi_1 \xi_2 \ldots \xi_{k-1}(1-\xi_k)|^s\Big) \tag{6.6}$$

$$\leqq \sup_{\xi_1 \ldots \xi_{k-1}} \mathscr{E}_p(|\cos ln\, \xi_1 \xi_2 \ldots \xi_{p-1}(1-\xi_p)|^s) \times \mathscr{E}\Big(\prod_{k=1}^{p-1} |\cos ln\, \xi_1 \xi_2 \ldots \xi_{k-1}(1-\xi_k)|^s\Big)$$

where $\mathscr{E}_p$ is the expectation when $\xi_1, \xi_2, \ldots \xi_{p-1}$ are fixed. Now

$$\mathscr{E}_p(|\cos ln\, \xi_1 \xi_2 \ldots \xi_{p-1}(1-\xi_p)|^s) \tag{6.7}$$

is the mean value of $|\cos y|^s$ on an interval of length

$$ln\, \xi_1 \xi_2 \ldots \xi_{p-1}(b_p - a_p).$$

According to (6.1), (6.2) and (6.5), this length exceeds 2π when n is large enough, and therefore (6.7) is majorized by

$$\varepsilon_s = \frac{1}{\pi} \int_0^{2\pi} |\cos y|^s \mathrm{d}y.$$

By a repetition of this argument we may estimate

$$\mathscr{E}\Big(\prod_{k=1}^{p-1} |\cos ln\, \xi_1 \xi_2 \ldots \xi_{k-1}(1-\xi_k)|^s\Big)$$

and we obtain

$$\mathscr{E}(|\hat{\mu}(n)|^s) \leqq \varepsilon_s^p. \tag{6.8}$$

When β is small enough ($\beta = \beta(s, \delta)$), (6.8) implies (6.4), and that proves the theorem.

As the reader sees, the method is simple and powerful. A slight modification gives the following: *if we have* (6.1) *and*

$$\underline{\lim}\, (\sqrt{k}(a_1 \ldots a_k)^{1/k}) > 0, \tag{6.9}$$

a.s. E is a M_0-set.

The interest of (6.9), compared to (6.2), is that we may choose the b_k tending to 0. Then E is a very thin set: its Hausdorff dimension is zero, and the set of all linear combinations of elements of E with integral coefficients has Lebesgue measure zero. Obviously E is not an independent set. Rudin's example of an independent M_0-set is obtained from such an E by a careful modification of the approximating sets E_n [15, p. 103].

The above results do not give examples of closed sets with equal Hausdorff and Fourier dimensions. In order to obtain such examples, Salem introduces a more complicated kind of sets: at each step of the construction, each interval is replaced by a number of equal intervals, and the dissection is the same for all intervals of the same step; in this way E_n is a union of equal intervals, and all intervals of E_n carry equal portions of E. Again E carries a natural measure $d\mu$. Given $\alpha \in]0, 1[$, it is possible to randomise the above construction in order that a.s. the Hausdorff dimension of E is α and a.s. $d\mu \in M_\alpha$ [15, p. 106].

7. The Levy Sets. Fourier Dimension

First let us recall a few definitions related to positive stochastic processes with independent increments [16, chap. 5].

We define the canonical Poisson measure $P(dt\, du)$ on the region $t > 0$, $u > 0$ of the (t, u)-plane in the following way: for each borelian set B, $P(B)$ is a Poisson random variable with parameter $|B|$, the Lebesgue measure of B, i.e.

$$\mathscr{E}(e^{iyP(B)}) = e^{-|B|(1-e^{iy})}; \tag{7.1}$$

for two disjoint borelian sets B_1 and B_2, $P(B_1)$ and $P(B_2)$ are independent random variables. A. s. P is a sum of countably many unit masses $\delta_{(\theta_n, \eta_n)}$, i.e.

$$P(B) = \text{number of } (\theta_n, \eta_n) \text{ in } B.$$

Now we suppose that $\lambda(u)$ is a positive decreasing function on $]0, \infty[$, such that

$$\int_1^\infty \lambda(u) \mathrm{d}u < \infty, \tag{7.2}$$

and we define

$$X(t) = \int_{0<\tau<t} \int_{u>0} \lambda(u) P(\mathrm{d}u \, \mathrm{d}\tau). \tag{7.3}$$

Formally, (7.1) and (7.3) give

$$\begin{cases} \mathscr{E}(e^{iyX(t)}) = e^{-t\psi(y)} \\ \psi(y) = \displaystyle\int_0^\infty (1-e^{i\lambda(u)y}) \mathrm{d}u \end{cases} \tag{7.4}$$

and (7.2) shows that (7.4) is meaningful, i.e. $X(t) < \infty$ a.s. Instead of (7.3), we may write

$$X(t) = \sum_{0<\theta_n<t} \lambda(\eta_n); \tag{7.5}$$

this means, a.s. $X(t)$ is an increasing right-continuous jump function, with a jump $\lambda(\eta_n)$ at θ_n, and $X(0) = 0$. Since $X(t)$ has independent increments on disjoint intervals, (7.4) gives

$$\mathscr{E}(e^{i(y_1X_1+y_2(X_2-X_1)+\ldots+y_n(X_n-X_{n-1}))}) = e^{-(t_1\psi_1+(t_2-t_1)\psi_2+\ldots+(t_n-t_{n-1})\psi_n)} \tag{7.6}$$

for any increasing positive sequence $t_1, t_2, \ldots t_n$ and any real sequence $y_1, y_2, \ldots y_n$, with $X_j = X(t_j)$ and $\psi_j = \psi(y_j)$.

We define the Lévy set E associated with $\lambda(u)$ as the closure of $X([0, 1])$ (the range of $X(t)$ on $[0, 1]$). Following [14], we shall prove the following theorem: *let* $h(y) = |\psi(y)|$ and let $\psi(y)$ be a decreasing function such that $\psi(y) = O(\psi(2y))$ and let $h(y) = |\psi(y)|$ and let $\psi(y)$ be a decreasing function such that $\psi(y) = O(\psi(2y))$ and

$$\lim_{y\to\infty} \psi(y) = \sqrt{\frac{h(y)}{\log y}} = \infty. \tag{7.7}$$

Then a.s. E is a $M_{(\varphi)}$-set.

The proof uses the image by X of the Lebesgue measure on $[0, 1]$, i.e. the measure $d\mu$ carried by E such that

$$\int_0^x d\mu = \sup(t|t \leqq 1 \quad \text{and} \quad X(t) \leqq x).$$

An alternative definition is

$$\int g\, d\mu = \int_0^1 g(X(t))dt$$

for any continuous function g on E. The Fourier transform of $d\mu$ is

$$\hat{\mu}(y) = \int e^{i\lambda y} d\mu(\lambda) = \int_0^1 e^{iyX(t)} dt.$$

As in Salem's method, we shall compute $\mathscr{E}(|\hat{\mu}(y)|^{2p})$. We have

$$\begin{aligned}
|\hat{\mu}(y)|^{2p} &= \int_0^1 \cdots \int_0^1 e^{iy(X(\tau_1)+X(\tau_2)+\ldots+X(\tau_p)-X(\tau'_1)-X(\tau'_2)-\ldots-X(\tau'_p))} \\
&\qquad\qquad \times d\tau_1 d\tau_2 \ldots d\tau_p d\tau'_1 d\tau'_2 \ldots d\tau'_p \\
&= (p!)^2 \int_{0\leqq\tau_1\leqq\tau_2\leqq\ldots\leqq\tau_p\leqq 1} \int_{0\leqq\tau'_1\leqq\tau'_2\ldots\leqq\tau'_p\leqq 1} \ldots \\
&= (p!)^2 \sum_{\{\varepsilon_j\}} \int_{0\leqq t_1\leqq t_2\ldots\leqq t_{2p}\leqq 1} e^{iy(\varepsilon_1 X(t_1)+\varepsilon_2 X(t_2)+\ldots+\varepsilon_{2p}X(t_{2p}))} \\
&\qquad\qquad \times dt_1 dt_2 \ldots dt_{2p}
\end{aligned}$$

where the sum is taken for all systems $\{\varepsilon_j\}$ such that $\varepsilon_j = \pm 1$ $(j = 1, 2, \ldots 2p)$ and $\varepsilon_1 + \varepsilon_2 + \ldots + \varepsilon_{2p} = 0$. Now we may use (7.6) with $n = 2p$,

$$y_{2p} = \varepsilon_{2p} y,\ y_{2p-1} = (\varepsilon_{2p}+\varepsilon_{2p-1})y, \ldots y_1 = (\varepsilon_1+\varepsilon_2+\ldots+\varepsilon_{2p})y = 0;$$

we obtain

$$\mathscr{E}(|\hat{\mu}(y)|^{2p})$$
$$= (p!)^2 \sum_{\{\varepsilon_j\}} \int_{0 \leqq t_1 \leqq t_2 \ldots \leqq t_{2p} \leqq 1} e^{-(t_1\psi_1 + (t_2 - t_1)\psi_2 + \ldots + (t_{2p} - t_{2p-1})\psi_{2p})} \times \mathrm{d}t_1 \, \mathrm{d}t_2 \ldots \mathrm{d}t_{2p}.$$

When j is even we have $|\psi_j| \geqq h(y)$ and anyway $\mathscr{R}\psi_j \geqq 0$. Therefore, whatever the ε_j may be, the last integral is bounded in absolute value by

$$\int_{0 \leqq t_1 \leqq t_3 \leqq \ldots \leqq t_{2p-1} \leqq 1} \left(\frac{2}{h(y)}\right)^p \mathrm{d}t_1 \, \mathrm{d}t_3 \ldots \mathrm{d}t_{2p-1} = \frac{2^p}{p!(h(y))^p}.$$

Therefore

$$\mathscr{E}(|\hat{\mu}(y)|^{2p}) \leqq \frac{2^p(2p)!}{p!(h(y))^p}.$$

Given a positive increasing sequence y_n, such that $\sum_1^\infty y^{-2} < \infty$, we define $p_n = [\log y_n]$. We have

$$\mathscr{E}\left(\sum_1^\infty y_n^{-2} \frac{|\hat{\mu}(y_n)|^{2p_n} p_n!(h(y_n))^{p_n}}{2^{p_n}(2p_n)!}\right) < \infty.$$

Therefore almost surely

$$\hat{\mu}(y_n) = O\left(\left(\frac{(2p_n)!\,y_n^2}{p_n!}\right)^{1/2p_n} \left(\frac{2}{h(y_n)}\right)^{\frac{1}{2}}\right) = o(\varphi(y_n)) \qquad (n \to \infty).$$

We may choose the sequence y_n in such a way that for each integer ν it contains all multiples of $2^{-\nu}$ except a finite number. Then we have

$$\hat{\mu}(n2^{-\nu}) = o(\varphi(n2^{-\nu})) \qquad (n \to \infty) \qquad (\nu = 1, 2, \ldots)$$

and, for the same reason as in (6.3), that implies

$$\hat{\mu}(y) = o(\varphi(y)) \qquad (y \to \infty).$$

Therefore a.s. E is a $M_{(\varphi)}$-set, as we had to prove.

There are two particular cases of interest, which we shall consider later.

1. If $\lambda(u) = u^{-1/\alpha}$ $(0 < \alpha < 1)$, $X(t)$ is called a stable process of index α. In this case it is pretty well known – and we shall prove in the next paragraph – that the Hausdorff dimension of E is α a.s. [4]. An obvious computation gives $\psi(y) = cy^\alpha$ (c complex). Therefore E is a $M_{(\varphi)}$-set with $\varphi(y) = y^{-\alpha/2} \log y$, a.s. As a consequence, the Fourier dimension of E is α a.s.*

* Let us add that these Lévy sets play an important role in the turbulence theory, and that [14] orginated from the ideas of B. Mandelbrot thereabout.

2. If $\lambda(y) = e^{-u^\alpha}_{(\alpha<1)}$, we have

$$\text{Re}\, \psi(y) \geqq (\log y)^{1/\alpha} + 0(1) \qquad (y \to \infty).$$

Therefore E is a M_0-set a.s. In this case we shall see that E is independent a.s.

8. The Levy Sets. Arithmetical Properties

There is no simple connexion between the Hausdorff dimension and the arithmetical properties of a set. But the Hausdorff dimension sits between the Fourier dimension and the Besicovitch-Taylor index, both of them having some arithmetical significance. Let us first define the Besicovitch-Taylor index of a compact set F on the line.

Let S be the smallest closed interval containing F. The complement of F with respect to S consists of countably many open intervals I_n (the end points of I_n belonging to F). We define $\beta(F)$ (the Besicovitch-Taylor index of F) as the infimum of the $\beta > 0$ such that

$$\sum |I_n|^\beta < \infty. \tag{8.1}$$

The connection with the Hausdorff dimension is the following: *if the Lebesgue measure of F is zero,*

$$\dim F \leqq \beta(F)[1]. \tag{8.2}$$

The proof is as follows. We may suppose $|I_{n+1}| \leqq |I_n|$, and write

$$r_n = \sum_{j=n}^{\infty} |I_j|, \qquad F_n = S - \bigcup_{j=n}^{n-1} I_j.$$

F_n consists of n closed intervals of total length r_n, $J_1^{(n)}, J_2^{(n)}, \ldots, J_n^{(n)}$, say. For any $\beta \varepsilon]0, 1]$, we have by a concavity argument

$$\sum_{i=1}^{n} |J_i^{(n)}|^\beta \leq n \left(\frac{r_n}{n}\right)^\beta = n^{1-\beta} r_n^\beta. \tag{8.3}$$

If (8.1) holds and $\beta \varepsilon]0, 1]$, we have $|I_n|^\beta = o(1/n)$ and therefore

$$r_n = o(n^{1-1/\beta}) \qquad (n \to \infty); \tag{8.4}$$

therefore F has Hausdorff measure 0 in the dimension β. Hence (8.2) holds.

Let us explain the arithmetical properties connected with $\beta(F)$ and $\mathscr{F}$-dim(F). Given F and a positive integer p, we write $\oplus_p F$ for the set of all real x that can be written in the form

$$x = x_1 + x_2 + \dots x_p \qquad (x_j \in F)$$

and $\boxplus_p F$ for the set of all real x that can be written in the form

$$x = q_1 x_1 + q_2 x_2 + \dots q_p x_p \qquad (x_j \in F,\ q_j \text{ rational}). \tag{8.5}$$

Obviously $\oplus_p F \subset \boxplus_p F$. Now we prove the following: *if the Lebesgue measure of F is zero and $p < \beta^{-1}(F)$, the Lebesgue measure of $\boxplus_p F$ is zero.*

The proof is almost the same as before. If (8.5) holds, $x_1, x_2, \dots x_p$ belong to p different intervals $J_\alpha^{(n)}$ when n is chosen large enough, therefore x belongs to

$$\bigcap_{n=1}^{\infty} \bigcup_{1 \leqq \alpha_1, \alpha_2, \dots \alpha_p \leqq n} (q_1 J_{\alpha_1}^{(n)} + q_2 J_{\alpha_2}^{(n)} + \dots + q_p J_{\alpha_p}^{(n)}). \tag{8.6}$$

We may suppose $|J_1^{(n)}| \geqq |J_2^{(n)}| \geqq \dots \geqq |J_n^{(n)}|$ and write $b = \sup(|q_1|, |q_2|, \dots |q_p|)$. Then

$$\text{meas}\,(q_1 J_{\alpha_1}^{(n)} + q_2 J_{\alpha_2}^{(n)} + \dots + q_p J_{\alpha_p}^{(n)}) \leqq pb |J_{\inf(\alpha_1, \alpha_2, \dots \alpha_p)}^{(n)}|.$$

Since $|J_1^{(n)}| + |J_2^{(n)}| + \dots |J_n^{(n)}| = r_n$, the Lebesgue measure of (8.6) is dominated by

$$\varliminf_{n \to \infty} (pbn^{p-1} r_n). \tag{8.7}$$

Now the assumption $p < \beta^{-1}(F)$ implies (8.1) with $\beta = p^{-1}$, and (8.1) implies (8.4). Therefore (8.7) vanishes. Therefore $\boxplus_p F$ is a countable union of sets of Lebesgue measure zero, and our assertion is proved. Moreover, we may replace the assumption $p < \beta^{-1}(F)$ by the weaker assumption

$$\sum |I_n|^{1/p} < \infty.$$

As a particular case, *if $\beta(F) = 0$, F does not generate the real line, considered as a linear space over the rationals.*

In the opposite direction, we may use the Fourier dimension of F. Actually, *if $p > (\mathscr{F} - \dim F)^{-1}$, the Lebesgue measure of $\oplus_p F$ is positive.*

The proof is very easy. For $\oplus_p F$ carries $*_p\, d\mu$ (the *p-th* power of convolution of $d\mu$) and the Fourier transform of $*_p \, d\mu$ is $\hat{\mu}^p$. The assumption implies $\hat{\mu}^p \in L^2$, therefore $*_p \, d\mu$ is absolutely continuous, therefore meas. $\oplus_p F > 0$.

In the same way, *if* $p > 2(\mathscr{F}\text{-dim } F)^{-1}$, $\oplus_p F$ *contains an interval.*

Instead of F, let us consider again the random Lévy set E associated with a positive decreasing function $\lambda(u)$. In addition to (7.2), let us suppose

$$\int_1^\infty \lambda^\beta(u)du < \infty, \qquad 0 < \beta \leqq 1. \tag{8.8}$$

Since (7.2) implies that (7.5) is bounded a.s., (8.8) implies

$$\sum_{0<\theta_n<1} \lambda^\beta(\eta_n) < \infty. \tag{8.9}$$

Now (8.9) is nothing but (8.1). Therefore, *if* (8.8) *holds, a. s.* dim $E \leqq \beta$. *Moreover, if* $p \leqq 1/\beta$, *a. s.* $\boxplus_p E$ *is a set of Lebesgue measure zero.*

As an application, let us consider a stable process of index α. In the last paragraph we proved $\mathscr{F}$-dim $E \geqq \alpha$ a.s., and since (8.8) holds whenever $\beta > \alpha$ we have $\beta(E) \leqq \alpha$ a.s. *Therefore*

$$\text{a.s. } \beta(E) = \mathscr{F}\text{-dim } E = \text{dim } E = \alpha.$$

Moreover, if $p < 1/\alpha$, *a.s.* $\oplus_p E$ *has Lebesgue measure zero*; *if* $p > 1/\alpha$, a. s. $\oplus_p E$ *has a positive Lebesgue measure.*

The case $p = 1/\alpha$ would need a more careful investigation.

Now we suppose that (8.8) *holds for all* $\beta > 0$. Then a.s. E does not generate the real line. We shall prove much more, namely that *E is an independent set* whenever $\lim_{y\to\infty} (\text{Re } \psi(y)/\log y) = \infty$ and $\lambda(u)$ decreases strictly.*

It is enough to prove the following: given p rational numbers $q_1, q_2, \ldots q_p$, different from 0, and $2p$ rational numbers $a_1, b_1, a_2, b_2, \ldots a_p, b_p$, such that

$$0 = a_1 < b_1 < a_2 < b_2 < \ldots < a_p < b_p = 1,$$

the event

$$\exists\, t_j \in]a_j, b_j] \quad (j = 1, 2, \ldots p) | \sum_{j=1}^p q_j X(t_{j\pm 0}) = 0 \tag{8.10}$$

* The necessity of the condition on $\lambda(u)$ was pointed out to me by Frode Poulsen.

has probability zero. Let us define $A = \bigcup_{j=1}^{p}]a_j, b_j]$, and

$$Y(t) = \sum_{0<\theta_n<t,\ \theta_n\in A} \lambda(\eta_n).$$

If we compare with (7.5), we see that the range of Y and the range of X are equally distributed. Therefore the set

$$E' = \{\sum_{j=1}^{p} q_j Y(t_j),\ t_j \in]a_j, b_j]\}$$

has measure zero a.s. Moreover, we have

$$X(t_j)-Y(t_j)$$
$$= X(a_2)-X(b_1)+X(a_3)-X(b_2)+\ \ldots\ +X(a_j)-X(b_{j-1}) = H_j$$

when $t_j \in]a_j, b_j]$. Then (8.10) implies

$$\sum_{j=1}^{p} q_j H_j \in E'. \tag{8.11}$$

For some q_j the first number is 0; then (8.11) has probability 0 because $\lambda(u)$ decreases strictly. If not, using (7.4) and the assumption on $\psi(y)$, the first member of (8.11) is seen to be a random variable with an absolutely continuous distribution, which depends only on the (θ_n, η_n) such that $\theta_n \notin A$. The second member, E', depends only on the (θ_n, η_n) such that $\theta_n \in A$. Therefore they are stochastically independent, and since E' has measure zero a.s., (8.11) has probability zero. That proves our theorem.

It is easy to find examples where (8.8) holds for all $\beta > 0$, and $\lim_{y\to\infty} (\operatorname{Re}\psi(y)/\log y) = \infty$. The corresponding Lévy sets give examples of independent M_0-sets. This answers a question which was put in [14].

If we choose $\lambda(u) = e^{-u^\alpha}$, $\alpha > 0$, the situation is the same if $\alpha \leqq 1$: the corresponding Lévy sets are a.s. independent M_0-sets. If $\alpha > 1$, they are independent again, but no longer M_0-sets. More generally, *if*

$$\int_1^\infty \left(\log \frac{1}{\lambda(u)}\right)^{-1} du < \infty,$$

a.s. E is not a M_0*-set.*

For the assumption implies easily $n = o(\log 1/r_n)$, a.s. Then we apply a theorem of Salem [15, p. 95]: a. s. E is a N-set, and it is known that no N-set is a M_0-set [15, p. 91].

That proves that $\sqrt{\log y}$ cannot be replaced by any $\log^{\alpha} y$, $\alpha < \frac{1}{2}$, in the definition of $\varphi(y)$ p. 86 (7.7).

9. Gaussian Fourier Series. Range

Let us consider a gaussian trigonometric series

$$\sum_{0}^{\infty} a_n(X_n \cos 2\pi nt + Y_n \sin 2\pi nt) \tag{9.1}$$

where $a_n \geqq 0$, and $X_0, X_1, Y_1, X_2, Y_2, \ldots$ are independent gaussian normal variables, i.e.

$$\mathscr{E}(e^{iyX_j}) = \mathscr{E}(e^{iyY_k}) = e^{-\frac{1}{2}y^2}.$$

In the case $a_0 = 1$, $a_1 = a_2 = \ldots = 2$, (9.1) is the Fourier series of the homogenous chaos on the circle (see p. 73), and the integrated series

$$X_0 t + \sum_{1}^{\infty} \frac{1}{\pi n}(Y_n \cos 2\pi nt - X_n \sin 2\pi nt) \tag{9.2}$$

represents the Wiener function of the brownian motion. All gaussian 1-periodic stationary processes are represented in the form (9.1).

We are interested in the case (9.1) is a.s. the Fourier series of a continuous function, and we denote this random function by $F(t)$. Its properties depend on the coefficients a_n. Roughly speaking, either $F(t)$ is tame a.s., or $F(t)$ is homogeneously wild a.s. For example, if $a_n = 0(n^{-\gamma})$, $\gamma > \frac{3}{2}$, a.s. F is continuously differentiable a.s.; in the opposite direction, if $a_n > n^{-\frac{3}{2}}$ $(n = 1, 2, \ldots)$, F is nowhere differentiable a.s. Many results of this type can be found in [13].

We introduce also a multidimensional series

$$\sum_{0}^{\infty} a_n(X_n \cos 2\pi nt + Y_n \sin 2\pi nt), \tag{9.3}$$

where $X_0, X_1, Y_1, \ldots$ are independent gaussian p-dimensional normal variables. $X_n = (X_n^1, X_n^2, \ldots X_n^p)$ where the X_n^j are independent real normal variables. In the case (9.3) represents a continuous function a.s., we denote it by $F(t)$. The range of F is in $\boldsymbol{R}^p$.

The main problem we shall consider in this section is the following: given a closed set E on the circle, what is the dimension of $F(E)$?

As a typical case, let us give the solution when $a_n = n^{-\gamma}$ ($\frac{1}{2} < \gamma < \frac{3}{2}$): then

$$\dim F(E) = \inf\left(p, \frac{2}{\gamma - 1} \dim E\right). \tag{9.4}$$

Roughly speaking, dim $F(E)$ is the largest dimension one can expect from the lipschitzian character of F and the dimension of E. We shall state and explain the results; for the detailed proofs, the reader is adviced to look at [13].

Many properties of (9.1) or (9.3) depend on the sequence

$$s_j = \Big(\sum_{2^j < n \leqq 2^{j+1}} a_n^2\Big)^{\frac{1}{2}}.$$

For example, a necessary condition for (9.1) or (9.3) to represent a continuous function is $\sum_1^\infty s_j < \infty$. A sufficient condition is $\sum_1^\infty s_j < \infty$ *and* s_j decreasing. Let us write

$$\sigma = \varliminf_{j\to\infty} \frac{-\log s_j}{j \log 2}, \qquad \tau = \varlimsup_{j\to\infty} \frac{-\log s_j}{j \log 2}.$$

σ is a critical value of the lipschitzian character of F. As usual, we write $f \in \Lambda_\alpha$ if either $0 < \alpha \leqq 1$ and

$$\sup_{t, t'} \frac{|f(t) - f(t')|}{|t - t'|^\alpha} < \infty$$

(in this case we say that f is lipschitzian or order α), or if $\alpha > 1$, f is differentiable and $f' \in \Lambda_{\alpha-1}$. We have a similar definition for a p-dimensional function f. *Then a.s.* $F \in \Lambda_\alpha$ *if* $\alpha > \sigma$, *and a.s.* $F \notin \Lambda_\alpha$ *if* $\alpha > \sigma$ [9].

Now, given a closed set E on the circle and a p-dimensional function f defined on the circle, $f \in \Lambda_\alpha$, $0 < \alpha < 1$, it is very easy to see that $\dim f(E) \leqq (1/\alpha) \dim E$. Therefore

$$\text{a.s.} \quad \dim F(E) \leqq \frac{1}{\sigma} \dim E \qquad \textit{if } \sigma < 1. \tag{9.5}$$

If $\sigma \geqq 1$, we have only $\dim F(E) \leqq \dim E$ a.s.

In the opposite direction, it is possible to prove

$$\text{a.s.} \quad \dim F(E) \geqq \inf\left(p, \frac{1}{\tau} \dim E\right). \tag{9.6}$$

In the case $a_n = n^{-\gamma}$ ($\frac{1}{2} < \gamma < \frac{3}{2}$), (9.5) and (9.6) give (9.4).

In order to prove (9.6), the idea is to consider a positive measure $\lambda(\mathrm{d}t)$ carried by E and its image by F, $\mu(\mathrm{d}x)$. Since

$$\int_{R^p} g(x)\mu(\mathrm{d}x) = \int_E g(F(t))\lambda(\mathrm{d}t)$$

for any continuous and bounded function g defined on R^p, we have

$$\hat{\mu}(u) = \int_{R^p} \mathrm{e}^{2\pi i(u,x)}\mu(\mathrm{d}x) = \int_E \mathrm{e}^{2\pi i(u,F(t))}\lambda(\mathrm{d}t)$$

($u \in R^p$, $(u, x) = u_1 x_1 + \ldots u_p x_p$) and an easy computation gives

$$\begin{cases} \mathscr{E}(|\hat{\mu}(u)|^2) = \iint \mathrm{e}^{-2\pi|u|^2\rho(t-t')}\lambda(\mathrm{d}t)\lambda(\mathrm{d}t') \\ \rho(t) = \sum_1^\infty a_n^2(1-\cos 2\pi nt) \end{cases} \tag{9.7}$$

If $0 < \beta < p$, the energy integral of $\mu(\mathrm{d}x)$ with respect to the kernel $|x|^{-\beta}$ is

$$\int_{R^p} |\hat{\mu}(u)|^2 |u|^{\beta-p}\mathrm{d}u.$$

According to (9.7), its expectation is

$$\iint (\rho(t-t'))^{-\frac{1}{2}\beta}\lambda(\mathrm{d}t)\lambda(\mathrm{d}t') \tag{9.8}$$

up to a constant factor. Therefore, if (9.8) is finite for some positive measure $\lambda(\mathrm{d}t)$ carried by E, a.s. the β-capacity of $F(E)$ is positive, and consequently $\dim F(E) \geqq \beta$.

An easy estimate gives

$$(\rho(t))^{-1} = 0(|t|^{-2(\tau+\varepsilon)}) \qquad (\varepsilon > 0).$$

Therefore (9.8) is finite when $\beta(\tau+\varepsilon) < 1$ and the energy intergal of $\lambda(\mathrm{d}t)$ with respect to the kernel $|t|^{-\beta(\tau+\varepsilon)}$ is finite. Therefore (9.8) is finite for some positive measure $\lambda(\mathrm{d}t)$ when $\beta(\tau+\varepsilon) < \dim E$. That is the way (9.6) is proved.

From (9.7) it is possible to obtain conditions on the a_n and E such that a.s. $F(E)$ has a positive p-dimensional measure, of $F(E)$ contains an open set. We leave them as exercises for the reader.

There is no direct use of the sets $F(E)$ in harmonic analysis. It is natural to ask whether the Fourier dimension and the Hausdorff dimension of these sets are the same. We suspect that the answer is positive for some sequences a_n, but we have no result in this direction.*

10. Gaussian Fourier Series. Zeros

Let us consider again a series (9.1), and suppose that a.s. it represents a continuous function $F(t)$. We are interested by the set of all zeros of F on the circle, $F^{-1}(0)$. This random set is interesting in itself. It is also interesting in connection with the so-called problem of spectral synthesis.

Either in order to study the properties or in order to make use use of $F^{-1}(0)$, we try to define measures or distributions carried by $F^{-1}(0)$. The most interesting measures or distributions of this type can be written as $\delta(F)$ or $\delta^{(q)}(F)$, where δ is the Dirac measure, and $\delta^{(q)}$ its q-th derivative. Of course, they need a careful definition.

We consider again, as in (9.7),

$$\rho(t) = \sum_{1}^{\infty} a_n^2(1-\cos 2\pi nt)$$

and we make the assumption

$$\int_0^1 \frac{dt}{t^\alpha\sqrt{\rho(t)}} < \infty, \qquad 0 < \alpha < 1. \tag{10.1}$$

Moreover, we assume $\sum_0^\infty a_n^2 = 1$ for conveniency; then $\mathscr{E}(e^{2\pi iuF(t)}) = e^{-\pi u^2}$. Let us define

$$T_\varepsilon(t) = \frac{1}{\sqrt{\varepsilon}}\,e^{-\pi/\varepsilon|F(t)|^2} = \int e^{-\pi\varepsilon u^2}e^{2\pi iuF(t)}\,du.$$

We may write $T_\varepsilon = \delta_\varepsilon(F)$, and δ_ε is an approximate Dirac measure. In order to define $T = \delta(F)$, we shall exhibit a Cauchy sequence T_{ε_n} in a convenient Hilbert space $\mathscr{H}$.

$\mathscr{H}$ is the space of all distributions on the circle of finite energy with respect to $\kappa(t) = |\mathrm{tg}\,\pi t|^{-\alpha}$:

$$\theta \sim \sum \hat{\theta}_n e^{2\pi int}$$

$$\|\theta\|_{\mathscr{H}}^2 = \sum |\hat{\theta}_n|^2\hat{\kappa}_n(\kappa(t) = \sum \hat{\kappa}_n e^{2\pi int}).$$

* Since this paper was written, we proved this conjecture: Compt. Rend. **263**, pp 678–681 (1966).

We have

$$||T_\varepsilon - T_\eta||^2_{\mathscr{H}} = \int_0^1 \int_0^1 (T_\varepsilon(t) - T_\eta(t))(T_\varepsilon(t') - T_\eta(t'))\kappa(t-t')\mathrm{d}t\,\mathrm{d}t'$$

and a standard computation gives

$$\mathscr{E}(||T_\varepsilon - T_\eta||^2_{\mathscr{H}}) = \int_0^1 \int_{-\infty}^{\infty} \int_{-\infty}^{\infty} (\mathrm{e}^{-\pi\varepsilon u^2} - \mathrm{e}^{-\pi\eta u^2})(\mathrm{e}^{-\pi\varepsilon u'^2} - \mathrm{e}^{-\pi\eta u'^2})\mathrm{e}^{-\pi(u-u')^2 + 2\pi u u'\rho(t)} \times \kappa(t)\mathrm{d}u\,\mathrm{d}u'\,\mathrm{d}t.$$

Because of the assumption (10.1), the integral

$$\int_0^1 \int_{-\infty}^{\infty} \int_{-\infty}^{\infty} \mathrm{e}^{-\pi 1(u-u')^2 + 2\pi u u'\rho(t)}\kappa(t)\mathrm{d}u\,\mathrm{d}u'\,\mathrm{d}t = \int_0^1 \frac{\kappa(t)\mathrm{d}t}{(1-(1-\rho(t)^2))^{\frac{1}{2}}}$$

is finite. By the Lebesgue theorem, we have

$$\lim_{\varepsilon,\eta\to 0} \mathscr{E}(||T_\varepsilon - T_\eta||^2_{\mathscr{H}}) = 0. \tag{10.2}$$

Choosing a sequence ε_n tending to zero such that

$$\mathscr{E}(||T_{\varepsilon_{n+1}} - T_{\varepsilon_n}||^2_{\mathscr{H}}) < n^{-4}$$

we have

$$\mathscr{E}(||T_{\varepsilon_{n+1}} - T_{\varepsilon_n}||_{\mathscr{H}}) < n^{-2}$$

$$\mathscr{E}(\sum_1^\infty ||T_{\varepsilon_{n+1}} - T_{\varepsilon_n}||_{\mathscr{H}}) < \infty$$

and therefore

$$\text{a.s.} \quad \sum_1^\infty ||T_{\varepsilon_{n+1}} - T_{\varepsilon_n}||_{\mathscr{H}} < \infty.$$

We define T as the limit of T_{ε_n} in $\mathscr{H}$. Because of (10.2) it does not depend on the sequence ε_n we choose. Since T_ε is a positive function, T is a measure $\geqq 0$. Moreover, since T_ε tends to zero out of any neighborhood of $F^{-1}(0)$, T is carried by $F^{-1}(0)$. We may write $T = \delta(F)$.

The main point is that T is not a.s. the zero measure. In order to prove it, we compute

$$\mathscr{E}\left(\int_0^1 T_\varepsilon(t)\mathrm{d}t\right) = \frac{1}{1+\varepsilon}$$

$$\mathscr{E}\left(\left(\int_0^1 T_\varepsilon(t)\mathrm{d}t\right)^2\right) \leqq \int_0^1 \frac{\mathrm{d}t}{(1-(1-\rho(t))^2)^{\frac{1}{2}}}$$

and we apply the formula

$$p(X > \lambda\mathscr{E}(X)) > (1-\lambda)^2 \frac{\mathscr{E}^2(X)}{\mathscr{E}(X^2)}$$

with $X = \int_0^1 T_\varepsilon(t)\mathrm{d}t$, $\lambda = \frac{1}{2}$. We obtain

$$\lim_{\varepsilon\to 0} p\left(\int_0^1 T_\varepsilon(t)\mathrm{d}t > \tfrac{1}{3}\right) > 0.$$

Since $\int_0^1 T_{\varepsilon_n}(t)\mathrm{d}t$ tends to the total mass of T as $n \to \infty$, we have $p(T \neq 0) > 0$.

As a consequence we have the following result: *if* (10.1) *holds*, $p(\dim F^{-1}(0) \geqq \alpha) > 0$. Let us remark that we do not have $p(\dim F^{-1}(0) \geqq \alpha) = 1$ when $a_0 \neq 0$ in (9.1), because in this case $p(F^{-1}(0) = \emptyset) > 0$. If $a_0 = 0$, it seems likely that $p(\dim F^{-1}(0) \geqq \alpha) = 1$, but we do not know how to prove it.

The preceding result can be obtained also for series (9.3) instead of (9.1); then (10.1) has to be replaced by

$$\int_0^1 \frac{\mathrm{d}t}{t^\alpha(\rho(t))^{p/2}} < \infty \qquad (0 < \alpha < 1).$$

It is possible also to consider the case $\alpha = 0$. Then the conclusion is that $F^{-1}(0)$ is not countable with a positive probability. For the details of computations and complements, the reader is referred to [13].

11. Gaussian Fourier Series and Spectral Synthesis

In order to define $\delta^{(q)}(F)$, where F is again a random continuous function represented by (9.1), we consider the integral

$$\int_{-\infty}^{\infty} (2\pi i u)^q \mathrm{e}^{2\pi i u F(t)}\mathrm{d}u. \tag{11.1}$$

As a space of distributions we consider $\mathscr{F}l^\gamma$ ($\gamma > 2$), i.e. the space of all distributions $\theta \sim \sum \hat{\theta}_n e^{2\pi i n t}$ such that

$$||\theta||_{\mathscr{F}l^\gamma} = (\sum |\hat{\theta}_n|^\gamma)^{1/\gamma} < \infty.$$

We are looking for a condition on the a_n such that the integral (11.1) is convergent in $\mathscr{F}l^\gamma$. More precisely, we want

$$\int_{-\infty}^{\infty} |u|^q \| e^{2\pi i u F(t)} \|_{\mathscr{F}l^\gamma} du < \infty \quad \text{a.s.}. \tag{11.2}$$

For the calculations, the most convenient value of γ is $\gamma = 4$, because

$$\begin{aligned}\xi(u) &= \mathscr{E}(\| e^{2\pi i u F(t)} \|_{\mathscr{F}l^4}^4) \\ &= \int_0^1 \int_0^1 \int_0^1 \mathscr{E}(e^{2\pi i u F(s) - F(s-t) - F(s') - F(s'-t)}) ds\, ds'\, dt\end{aligned}$$

can be computed rather easily. One finds

$$\xi(u) = \int_0^1 \int_0^1 \int_0^1 e^{-16\pi u^2 \varphi(t, s-s')} ds\, ds'\, dt$$

where

$$\varphi(t, s) = \sum_1^\infty a_n^2 \sin^2 \frac{ns}{2} \sin^2 \frac{nt}{2}. \tag{11.3}$$

Given $r > 1$, we have

$$\int_0^\infty u^{2r-1} \xi(u) du = \tfrac{1}{2}\Gamma(r) \int_0^1 \int_0^1 (16\pi\varphi(t, s))^{-r} ds\, dt.$$

Therefore the condition

$$\int_0^1 \int_0^1 (\varphi(t, s))^{-r} ds\, dt < \infty \tag{11.4}$$

implies that

$$\text{a.s.} \quad \int_0^\infty u^{2r-1} \| e^{2\pi i u F(t)} \|_{\mathscr{F}l^4}^4 du < \infty.$$

By the Hölder inequality, this implies (11.2) when $\gamma = 4$ and $r > 2(q+2)$. Then it is easy to check that a.s. (11.1) defines a distribution $T_q \in \mathscr{F}l^4$, carried by $F^{-1}(0)$; actually, we have $T_q = \lim_{\varepsilon \to 0} \delta_\varepsilon^{(q)}(F)$ where $\delta_\varepsilon(x)$ is the approximate Dirac measure that was introduced in the last section.

Therefore, if we have (11.4) *and* $r > 2(q+2)$, (11.1) *represents a.s. a distribution* $T_q \in \mathscr{F}l^4$, *carried by* $F^{-1}(0)$.

As an example, we may take $a_n = 4^{-j\eta}$ when $n = 4^{-j}$ and $a_n = 0$ when n is not any power of 4 ($\eta > 0$). Then (11.4) holds whenever η is small enough, namely

$$\sum (\tfrac{3}{4})^j 4^{2\eta r j} < \infty;$$

we omit the proof, a simple computation which can be found in [8], [10] or [15, chap. IX].

We have not yet proved that T_q is non zero with a positive probability. Actually we shall prove the following theorem, which solves the question: *if we have* (11.4), $r > 2(q+2)$, *and* $\sum_0^\infty a_n < \infty$, *then*

$$p((T_q, F^q) \neq 0) > 0. \tag{11.5}$$

Here (T_q, F^q) is given by a formal Parceval relation

$$\sum_{-\infty}^{\infty} \hat{T}_q(n)(\hat{F}^q)(-n).$$

The proof will consist in computing $\mathscr{E}((T_q, F^q))$ and check that it does not vanish. Since (11.2) holds with $\gamma = 4$ and a fortiori with $\gamma = \infty$, and a.s. $F^q \in \mathscr{F}l^1$, we have

$$(T_q, F^q) = \int_{-\infty}^{\infty} \left(\int_0^1 F^q(t)(2\pi i u)^q e^{2\pi i u F(t)} \mathrm{d}t \right) \mathrm{d}u$$

and a.s. the integral is convergent. Integrating by parts we have

$$(T_q, F^q) = \frac{(-1)^q}{q!} \int_{-\infty}^{\infty} \left(\int_0^1 e^{2\pi i u F(t)} \mathrm{d}t \right) \mathrm{d}u$$

and therefore

$$\mathscr{E}(T_q, F^q) = \frac{(-1)^q}{q!} \int_{-\infty}^{\infty} e^{-\pi u^2} \mathrm{d}u \qquad \left(\sum_0^\infty a_n^2 = 1\right)$$
$$= \frac{(-1)^q}{q!}.$$

That proves (11.5).

Let us consider the case $q = 1$. Then (11.5) expresses that there exist $F \in \mathscr{F}l^1$ and $T_1 \in \mathscr{F}l^\infty$, carried by $F^{-1}(0)$, such that $(T_1, F) \neq 0$. That is the probabilistic proof of Malliavin's theorem on spectral synthesis. T_1 *does not* belong to the weak closure of the measures carried by its support in $\mathscr{F}l^\infty$ (considered as dual of $\mathscr{F}l^1$). *F cannot* be approximated in $\mathscr{F}l^1$ by means of functions which vanish in a neighborhood of $F^{-1}(0)$. $F^{-1}(0)$ *is not* a set of spectral synthesis, i.e. the ideal of functions in $\mathscr{F}l^1$ which vanish on $F^{-1}(0)$ is not the closed ideal generated by the functions which vanish in a neighbourhood of $F^{-1}(0)$.

Though the probabilistic proof of Malliavin's theorem expresses that 'almost all' closed sets are sets of non spectral synthesis, no explicit geometrical construction of such a set is known until now.

References

[1] Besicovitch A. S. and Taylor S. J., On the complementary intervals of a linear closed set of zero Lebesgue measure, *J. London Math. Soc.* 29 (1954), p. 449.

[2] Billard, P., Séries de Fourier aléatoirement bornées, ... *Studia Math.* 22 (1963), pp. 310–330.

[3] Billard, P., Séries de Fourier aléatoirement bornées, continués, uniformément convergentes, *Annales Sc. E.N.S.* 82 (1965), pp. 131–179.

[4] Blumenthal, R. M. and Getoor, R. K., Some theorems on stable processes, *Trans. Amer. Math. Soc.* 95 (1960), pp. 263–273.

[5] Helson, H., Kahane, J.-P., A Fourier Method in diophantine problems, *J. Anal. Math.* XV (1965), 245–262.

[6] Ivasev-Musatov, O. S., (О. С. Ивашев-Мусатов), On coefficients of trigonometric null-series (о коэффициентах тригонометрических нуль-рчуов), *Izv. Akad. Nauk S.S.S.R.* (ser. Mat). 21 (1957), 559–578.

[7] Kahane, J.-P., Sur le recouvrement d'un cercle par des arcs disposés au hasard, *C. R. Acad. Sci. Paris* 248 (1959), pp. 184–186.

[8] Kahane, J.-P., Sur un théorème de Paul Malliavin, *C. R. Acad. Sci. Paris* 248 (1959) pp. 2943–2944.

[9] Kahane, J.-P., Propriétés locales des fonctions à séries de Fourier aléatoires, *Studia Math.* 19 (1960), pp. 1–25.

[10] Kahane, J.-P., Sur la synthèse harmonique dans 1^{∞}, *Anais da academia brasileira de Cièencias*, 32 (1960), pp. 179–189.

[11] Kahane, J.-P., Sur la divergence presque sûre presque partout de certaines séries de Fourier aléatoires, *Annales Universitatis Scientiarum Budapestinensis de Rolando Eötvös Nominatae*, Tomus III-IV, (1960/61), pp. 101–108.

[12] Kahane, J.-P., Pseudo-périodicité et séries de Fourier lacunaires, *Annales Sci. E.N.S.* 79 (1962), 93–150.

[13] Kahane, J.-P., Some Random Series of Functions, Heath, 1968.

[14] Kahane, J.-P., et Mandelbrot, B., Ensembles de multiplicité aléatoires, *C. R. Acad. Sci. Paris*, 26 (1965), 3931–3933.

[15] Kahane, J.-P., et Salem, R., *Ensembles parfaits et Séries trigonométriques*, Hermann, 1963.

[16] Lévy, P., *Processus stochastiques et mouvement brownien*, Paris 1948.

[17] Paley, R. E. A. C. and Zygmund, A., On some series of functions, *Proc. Cambridge Phil. Soc.* 26 (1930), 337–357 and 458–474; 28 (1932, 90–205.

[18] Paley, R. E. A. C. and Zygmund, A., A note on analytic functions in the unit circle, *Proc. Cambr. ph. Soc.* 28 (1932), 266–272.

[19] Salem, R., On sets of multiplicity for trigonometrical series, *Amer. J. Math.* 64 (1942), pp. 531–538.

[20] Salem, R., On singular monotonic functions whose spectrum has a given *H* dimension, *Arkiv för Mat.* 1 (1950), pp. 353–365.

[21] Varopoulos, N., Sur les mesures de Radon d'un groupe localement compact abélien, *C. R. Acad. Sci. Paris* 258 (1964), pp. 3805–3808.

[22] Wiener, N., Generalized Harmonic Analysis, *Acta Mathematica* 55 (1930), 117–258.

[23] Zygmund, A., *Trigonometric series*, Cambridge 1959.

Recent developments in the theory of finite Toeplitz operators

by I. I. Hirschman, Jr.

DEPARTMENT OF MATHEMATICS
WASHINGTON UNIVERSITY
ST. LOUIS, MISSOURI

Preface

The role in mathematics played by convolution operators

$$Wf \cdot (k) = \sum_{j=-\infty}^{\infty} c(k-j)f(j) \qquad -\infty < k < \infty$$

and their generalizations can scarcely be overstated. In particular these operators are fundamental to harmonic analysis on locally compact groups.

For some decades studies have been made of truncations of convolution operators such as

$$W^{+}f \cdot (k) = \sum_{j=0}^{\infty} c(k-j)f(j) \qquad 0 \leqq k < \infty,$$

and

$$W^{-}f \cdot (k) = \sum_{j=-\infty}^{0} c(k-j)f(j) \qquad -\infty < k \leqq 0.$$

The theory of these Toeplitz or Wiener-Hopf operators turns upon the fact that the linear order structure of the integers is compatible with the group structure of the integers and with their invariant measure. To some extent the integers may be replaced by an arbitrary Abelian group on which a compatible linear order structure has been distinguished. However, because many of the deeper results depend upon analytic function theory, these operators have been studied principally in the cases when the underlying group is either the integers or the real numbers. This theory, which is still under intensive development, has many applications. One, which it is particularly appropriate to mention here, is fluctuation theory in probability.

The study of finite section Toeplitz or Wiener-Hopf operators

$$W^{+}(n)f(k) = \sum_{j=0}^{n} c(k-j)f(j) \qquad 0 \leqq k \leqq n,$$

$$W^{-}(n)f(k) = \sum_{j=-n}^{0} c(k-j)f(j) \qquad -n \leqq k \leqq 0,$$

also goes back several decades, since in particular such operators appear in the definition of Szegö polynomials. Besides their applications to the theory of orthogonal polynomials they have important applications to the mathematics of computation, to certain topics in the theory of random walks, and also (somewhat less expectedly) to various problems in statistical mechanics. Although these operators have long been studied, it has only recently become apparent what very natural mathematical objects they are. In the present paper I have tried to give an introduction to the theory of finite Toeplitz operators from a point of view suggested by recent developments. I have also tried to describe, or at least mention, most of the recent work in this area. However, I have necessarily omitted much classical material, and for this the reader is referred to the books of Szegö [65], Grenander and Szegö [24], and Geronimus [23].

Since this paper was written, many interesting papers related to the subject matter of this article have appeared. I have not attempted to bring the bibliography up to date, but I feel that because of its particular relevance and importance I should mention "Projection Methods for Solving Wiener Hopf Equations" by I. A. Feldman and I. C. Gohberg, Moldavian Academy of Sciences, 1967, 163 pp. (in Russian).

1. Algebraic Theory of Finite Toeplitz Operators

1.1. Preliminaries

Let $\{c(k)\}_{-\infty}^{\infty}$ be a given sequence of complex numbers. In this chapter we will study the theory of the Toeplitz operators

$$W^{+}(n)f \cdot (k) = \sum_{j=0}^{n} c(k-j)f(j) \qquad k = 0, 1, \ldots, n, \tag{1.1.1}$$

and

$$W^{-}(n)f \cdot (k) = \sum_{j=-n}^{0} c(k-j)f(j) \qquad k = -n, \ldots, -1, 0 \quad (1.1.1')$$

for finite values of n; that is, we will not be concerned here (as we will in later chapters) with the asymptotic behaviour of $W^{+}(n)$ and $W^{-}(n)$

as $n \to \infty$. Although this theory is elementary in character, it contains a variety of interesting and important results.

As one would expect, Fourier series play an important role throughout. In later chapters when we investigate the behaviour of $\boldsymbol{W}^+(n)$ and $\boldsymbol{W}^-(n)$ as $n \to \infty$ the $\{\boldsymbol{c}(k)\}_{-\infty}^{\infty}$ will appear as Fourier coefficients of functions or measures. However, here the $\{\boldsymbol{c}(k)\}_{-\infty}^{\infty}$ are quite arbitrary. For this reason it is necessary for us to introduce generalized functions defined by formal Fourier series.

Let $\boldsymbol{Q}$ be the vector space of all complex valued functions on the integers I. If $\boldsymbol{f}$ and $\boldsymbol{g}$ are two such functions and if a and b are complex numbers we set

$$(a\boldsymbol{f}+b\boldsymbol{g})(k) = a\boldsymbol{f}(k)+b\boldsymbol{g}(k) \qquad -\infty < k < \infty.$$

With this definition $\boldsymbol{Q}$ is a vector space over the complex numbers. If $\boldsymbol{f}$ and $\boldsymbol{p} \in \boldsymbol{Q}$ and if $\boldsymbol{p}$ has finite support – that is, if there are finite integers $n_1 \leqq n_2$ such that $\boldsymbol{p}(k) = 0$ for $k < n_1$ and $k > n_2$ – we define the convolution $\boldsymbol{f} * \boldsymbol{p}$ in $\boldsymbol{Q}$ by

$$(\boldsymbol{f} * \boldsymbol{p})(k) = \sum_{j=n_1}^{n_2} \boldsymbol{f}(k-j)\boldsymbol{p}(j) \qquad -\infty < k < \infty.$$

It would be natural to define the Fourier transform $\mathscr{F}\boldsymbol{f} = f$ by the formula,

$$f(e^{i\theta}) = \sum_{k=-\infty}^{\infty} \boldsymbol{f}(k)e^{ik\theta} \qquad \theta \in T,$$

(where T is the real numbers modulo 2π), however, since the $\boldsymbol{f}(k)$'s are arbitrary, the series here will not in general converge. To avoid this difficulty we regard $f(e^{i\theta})$ as the name of this formal series and say that such f's are generalized functions on T. The space of all such generalized functions will be denoted by $\mathscr{Q}$. For $f = \mathscr{F}\boldsymbol{f}$ and $g = \mathscr{F}\boldsymbol{g}$ we define

$$\begin{aligned}(af+bg)\cdot(e^{i\theta}) &= \mathscr{F}[a\boldsymbol{f}+b\boldsymbol{g}] \\ &= \sum_{k=-\infty}^{\infty} [a\boldsymbol{f}(k)+b\boldsymbol{g}(k)]e^{ik\theta}.\end{aligned}$$

Similarly if $f = \mathscr{F}\boldsymbol{f}$, $p = \mathscr{F}\boldsymbol{p}$ where $\boldsymbol{p}$ has compact support we set

$$\begin{aligned}(fp)(e^{i\theta}) &= \mathscr{F}(\boldsymbol{f} * \boldsymbol{g}) && \theta \in T, \\ &= \sum_{k=-\infty}^{\infty} [\sum_{j=n_1}^{n_2} \boldsymbol{f}(k-j)\boldsymbol{p}(j)]e^{ik\theta} && \theta \in T.\end{aligned}$$

Finally we *define* a formal integral on $\mathscr{Q}$ by

$$\frac{1}{2\pi}\int_T f(e^{i\theta})d\theta = \boldsymbol{f}(0).$$

The inverse Fourier transform, $\boldsymbol{f} = \mathscr{F}^{-1}f$, is then given by the formula

$$\boldsymbol{f}(k) = \frac{1}{2\pi}\int_T f(e^{i\theta})e^{-ik\theta}d\theta.$$

Let $\boldsymbol{E}^+(n)$ and $\boldsymbol{E}^-(n)$ be linear operators on $\boldsymbol{Q}$ defined by the formulas

$$\boldsymbol{E}^+(n)\boldsymbol{f}\cdot(k) = \begin{cases}\boldsymbol{f}(k) & k \geqq n \\ 0 & k < n\end{cases},$$

$$\boldsymbol{E}^-(n)\boldsymbol{f}\cdot(k) = \begin{cases}0 & k > n \\ \boldsymbol{f}(k) & k \leqq n\end{cases}.$$

We *transfer* these operators to $\mathscr{Q}$ by putting

$$E^+(n)f\cdot(e^{i\theta}) = \mathscr{F}\boldsymbol{E}^+(n)\mathscr{F}^{-1}f, \qquad E^-(n)f\cdot(e^{i\theta}) = \mathscr{F}\boldsymbol{E}^-(n)\mathscr{F}^{-1}f;$$

thus

$$E^+(n)f\cdot(e^{i\theta}) = \sum_{k\geqq n}\boldsymbol{f}(k)e^{ik\theta},$$
$$E^-(n)f\cdot(e^{i\theta}) = \sum_{k\leqq n}\boldsymbol{f}(k)e^{ik\theta}.$$

These operators satisfy the relations:

$$\begin{aligned}
&E^+(n)E^-(m) = E^-(m)E^+(n) && \text{all } m, n,\\
&E^+(n)E^-(m)f = 0 && m < n,\\
&[E^+(n)+E^-(n-1)]f = f && \text{all } n,\\
&E^+(n)E^+(m) = E^+(\max[m, n]) && \text{all } m, n,\\
&E^-(n)E^-(m) = E^-(\min[m, n]) && \text{all } m, n,\\
&e^{im\theta}E^+(n)e^{-im\theta}f = E^+(n+m)f && \text{all } m, n,\\
&e^{im\theta}E^-(n)e^{-im\theta}f = E^-(n+m)f && \text{all } m, n.
\end{aligned} \tag{1.1.2}$$

The following notations will be useful. We denote by $\boldsymbol{Q}[a, b]$ the complex vector space of all functions $\boldsymbol{f}(k)$ where $a \leqq k \leqq b$. By setting $\boldsymbol{f}(k) = 0$ for all other values of k, $\boldsymbol{Q}[a, b]$ may be regarded as a subspace of $\boldsymbol{Q}$. $\mathscr{Q}[a, b]$ is defined to be the subspace of $\mathscr{Q}$ consisting of those $f \in \mathscr{Q}$ for which $\mathscr{F}^{-1}f \in \boldsymbol{Q}[a, b]$.

Let us now analyze the formula

$$\boldsymbol{W}^+(n)\boldsymbol{f}.(k) = \boldsymbol{g}(k) \qquad \boldsymbol{f}, \boldsymbol{g} \in \boldsymbol{Q}[0, n].$$

This is equivalent to the relation

$$c(e^{i\theta})f(e^{i\theta}) = g_1(e^{i\theta})+g(e^{i\theta})+g_2(e^{i\theta})$$

where $g_1 \in \mathcal{Q}[-\infty, -1]$ and $g_2 \in \mathcal{Q}[n+1, \infty]$. f and g, of course, belong to $\mathcal{Q}[0, n]$. This in turn is equivalent to

$$E^+(0)E^-(n)cf = g \qquad f, g \in \mathcal{Q}[0, n].$$

Similarly

$$\mathbf{W}^-(n)\mathbf{f}\,.\,(k) = \mathbf{g}(k) \qquad \mathbf{f}, \mathbf{g} \in \mathbf{Q}[-n, 0]$$

is equivalent to

$$E^-(0)E^+(-n)cf = g \qquad f, g \in \mathcal{Q}[-n, 0].$$

Thus if we define

$$\begin{aligned} W^+(n)f &= E^+(0)E^-(n)cf \qquad && f \in \mathcal{Q}[0, n], \\ W^-(n)f &= E^-(0)E^+(-n)cf && f \in \mathcal{Q}[-n, 0], \end{aligned}$$

$\mathbf{W}^+(n)$ and $W^+(n)$, and similarly $\mathbf{W}^-(n)$ and $W^-(n)$ are isomorphic operators.

1.2. Szegö polynomials

As in 1.1 we assume as given once and for all a sequence $\{\mathbf{c}(k)\}_{-\infty}^{\infty}$ of complex numbers. Let

$$D(n) = \det\,[\mathbf{c}(k-j)]_{k,\,j=0,\,\ldots,\,n}. \qquad n = 0, 1, \ldots. \tag{1.2.1}$$

Fix n. We *assume* $D(n) \neq 0$ so that $\mathbf{W}^+(n)$ and $\mathbf{W}^-(n)$ are invertible and we define $u(n, e^{i\theta})$ and $v(n, e^{i\theta})$ by

$$\begin{aligned} W^+(n)u(n) &= 1 \qquad && u(n) \in \mathcal{Q}[0, n], \\ W^-(n)v(n) &= 1 && v(n) \in \mathcal{Q}[-n, 0]. \end{aligned} \tag{1.2.2}$$

Let

$$u(n, e^{i\theta}) = \sum_{j=0}^{n} \mathbf{u}(n, j)e^{ij\theta}, \qquad v(n, e^{i\theta}) = \sum_{j=-n}^{0} \mathbf{v}(n, j)e^{ij\theta}.$$

Note that (1.2.2) is equivalent to

$$\begin{aligned} \sum_{j=0}^{n} \mathbf{c}(k-j)\mathbf{u}(n, j) &= \delta(k, 0) \qquad && k = 0, \ldots, n, \\ \sum_{j=-n}^{0} \mathbf{c}(k-j)\mathbf{v}(n, j) &= \delta(k, 0) && k = -n, \ldots, 0. \end{aligned} \tag{1.2.3}$$

The functions $u(n)$ and $v(n)$ play a basic role in all that follows. In this section we will use them to construct the Szegö polynomials. The values of $\boldsymbol{u}(n, j)$ and $\boldsymbol{v}(n, j)$ can be obtained from (1.2.3) by Cramer's rule. In particular we find that

$$\boldsymbol{u}(n, 0) = \boldsymbol{v}(n, 0) = D(n-1)/D(n).$$

Let us now *assume* further that $D(n-1) \neq 0$. Choose $d(n)$ so that

$$d(n)^2 = D(n-1)/D(n).$$

Theorem 2a. If $D(n)$, $D(n-1) \neq 0$ and if we set

$$\varphi(n, \mathrm{e}^{i\theta}) = d(n)^{-1}\mathrm{e}^{in\theta}v(n, \mathrm{e}^{i\theta}),$$
$$\psi(n, \mathrm{e}^{i\theta}) = d(n)^{-1}\mathrm{e}^{-in\theta}u(\mathrm{e}^{i\theta}),$$

then $\varphi(n) \in \mathscr{Q}[0, n]$, $\psi(n) \in \mathscr{Q}[-n, 0]$,

$$\left.\begin{aligned} \frac{1}{2\pi}\int_T \varphi(n, \mathrm{e}^{i\theta})\mathrm{e}^{-ik\theta}c(\mathrm{e}^{i\theta})d\theta &= 0 \\ \frac{1}{2\pi}\int_T \psi(n, \mathrm{e}^{i\theta})\mathrm{e}^{ik\theta}c(\mathrm{e}^{i\theta})d\theta &= 0 \end{aligned}\right\} \quad k = 0, 1, \ldots, n-1, \tag{1.2.4}$$

and

$$\frac{1}{2\pi}\int_T \varphi(n, \mathrm{e}^{i\theta})\psi(n, \mathrm{e}^{i\theta})c(\mathrm{e}^{i\theta})d\theta = 1. \tag{1.2.5}$$

Proof. From

$$c(\mathrm{e}^{i\theta})v(n, \mathrm{e}^{i\theta}) = (-\infty, -n-1)+1+(1, \infty)$$

(here $(1, \infty)$ stands for some function in $\mathscr{Q}[1, \infty]$, etc.) we see that

$$c(\mathrm{e}^{i\theta})\varphi(n, \mathrm{e}^{i\theta}) = (-\infty, -1)+d(n)^{-1}\mathrm{e}^{in\theta}+(n+1, \infty). \tag{1.2.6}$$

Similarly

$$c(\mathrm{e}^{i\theta})\psi(n, \mathrm{e}^{i\theta}) = (-\infty, -n-1)+d(n)^{-1}\mathrm{e}^{-in\theta}+(1, \infty). \tag{1.2.6$'$}$$

These relations imply (1.2.4) and also (1.2.5) if we note that $\varphi(n, n) = \psi(n, -n) = d(n)$.

$\varphi(n, e^{i\theta})$ and $\psi(n, e^{i\theta})$ are the biorthogonal Szegö polynomials introduced by Baxter [5]. Note that if [1] $c(-k) = c(k)^*$ then $\psi(n) = \varphi(n)^*$.

[1] For any complex number z, z^* is the conjugate of z.

Suppose that $D(k) \neq 0$, $k = 0, \ldots, n$. Then $\varphi(k, e^{i\theta})$ and $\psi(k, e^{i\theta})$ are defined for $k = 0, 1, \ldots, n$. Since

$$\varphi(k, e^{i\theta}) = \sum_{j=0}^{k} \varphi(k, j)e^{ij\theta} \qquad \varphi(k, k) \neq 0,$$

$$\psi(k, e^{i\theta}) = \sum_{j=-k}^{0} \psi(k, j)e^{ij\theta} \qquad \psi(k, -k) \neq 0,$$

it is evident that any $f \in \mathscr{Q}[0, n]$ can be written in the form

$$f(e^{i\theta}) = \sum_{k=0}^{n} f^{\sim}(k)\varphi(k, e^{i\theta}).$$

Similarly any $g \in \mathscr{Q}[-n, 0]$ can be written in the form

$$g(e^{i\theta}) = \sum_{k=0}^{n} g^{\sim}(k)\psi(k, e^{i\theta}).$$

It follows from Theorem 2a that the coefficients in these expansions can be computed as follows:

$$\begin{aligned} f^{\sim}(k) &= \frac{1}{2\pi}\int_T f(e^{i\theta})\psi(k, e^{i\theta})c(e^{i\theta})d\theta, \\ g^{\sim}(k) &= \frac{1}{2\pi}\int_T g(e^{i\theta})\varphi(k, e^{i\theta})c(e^{i\theta})d\theta. \end{aligned} \tag{1.2.7}$$

More generally if $D(k) \neq 0$, $k = n_1, n_1+1, \ldots, n_2$ and if $f \in \mathscr{Q}[0, n_2]$ then for each n, $n_1 \leqq n \leqq n_2$,

$$f(e^{i\theta}) = f_n(e^{i\theta}) + \sum_{k=n+1}^{n_2} f^{\sim}(k)\varphi(k, e^{i\theta})$$

where $f_n \in \mathscr{Q}[0, n]$ and where $f^{\sim}(k)$ is given by (1.2.7).

1.3. Recursion formulas

Theorem 3a. If $D(n), D(n-1) \neq 0$ then there exist (unique) constants $\alpha(n)$ and $\beta(n)$ such that

$$\begin{aligned} u(n, e^{i\theta}) &= u(n-1, e^{i\theta}) + \alpha(n)e^{in\theta}v(n, e^{i\theta}), \\ v(n, e^{i\theta}) &= v(n-1, e^{i\theta}) + \beta(n)e^{-in\theta}u(n, e^{i\theta}). \end{aligned} \tag{1.3.1}$$

Proof. The assumption $D(n), D(n-1) \neq 0$ ensures that $u(n)$, $v(n)$, $u(n-1)$, and $v(n-1)$ are defined. We have

$$W^+(n)u(n, \mathrm{e}^{i\theta}) = 1,$$

$$W^+(n)\mathrm{e}^{in\theta}v(n, \mathrm{e}^{i\theta}) = \mathrm{e}^{in\theta},$$

and

$$W^+(n)u(n-1, \mathrm{e}^{i\theta}) = 1-\alpha(n)\mathrm{e}^{in\theta},$$

for some constant $\alpha(n)$. Since $W^+(n)$ is invertible the first formula of (1) follows, etc.

Corollary 3b. If $D(n), D(n-1) \neq 0$ then

$$\begin{aligned}[1-\alpha(n)\beta(n)]u(n, \mathrm{e}^{i\theta}) &= u(n-1, \mathrm{e}^{i\theta})+\alpha(n)\mathrm{e}^{in\theta}v(n-1, \mathrm{e}^{i\theta}),\\ [1-\alpha(n)\beta(n)]v(n, \mathrm{e}^{i\theta}) &= v(n-1, \mathrm{e}^{i\theta})+\beta(n)\mathrm{e}^{-in\theta}u(n-1, \mathrm{e}^{i\theta}).\end{aligned} \tag{1.3.2}$$

If in addition $D(n-2) \neq 0$ then $\alpha(n)\beta(n) \neq 1$.

Proof. To obtain the first relation in (1.3.2) multiply the second relation of (1.3.1) by $\alpha(n)e^{in\theta}$ and add it to the first, etc. If $\alpha(n)\beta(n) = 1$ then

$$u(n-1, \mathrm{e}^{i\theta}) = -\alpha(n)\mathrm{e}^{in\theta}v(n-1, \mathrm{e}^{i\theta}).$$

This, however, is impossible since the coefficient of $e^{in\theta}$ is $-\alpha(n)\mathrm{d}(n-1)^2$ on the right, but is 0 on the left.

Let us set

$$G(n) = D(n)/D(n-1).$$

Corollary 3c. If $D(n), D(n-1), D(n-2) \neq 0$ then

$$[1-\alpha(n)\beta(n)]G(n-1) = G(n). \tag{1.3.3}$$

If $D(k) \neq 0$, $k = 0, 1, \ldots, n$, then

$$G(n) = c(0)\prod_{k=1}^{n}[1-\alpha(k)\beta(k)], \tag{1.3.4}$$

$$D(n) = c(0)^{n+1}\prod_{k=1}^{n}[1-\alpha(k)\beta(k)]^{n+1-k}. \tag{1.3.5}$$

Proof. If in either of the formulas (1.3.2) we compare the coefficients of 1 on the two sides we obtain (1.3.3). The formulas (1.3.4) and (1.3.5) are clearly consequences of (1.3.3).

Suppose that $D(n) \neq 0$, $n = n_1, n_1+1, \ldots, n_2$. If we write out the first formula of (1.3.1) for $n = n_1+1, \ldots, n_2$ and add we obtain

$$u(n_2, e^{i\theta}) = u(n_1, e^{i\theta}) + \sum_{n=n_1+1}^{n_2} \alpha(n)d(n)\varphi(n, e^{i\theta}). \tag{1.3.6}$$

Similarly

$$v(n_2, e^{i\theta}) = v(n_1, e^{i\theta}) + \sum_{n=n_1+1}^{n_2} \beta(n)d(n)\psi(n, e^{i\theta}). \tag{1.3.6'}$$

The relations (1.3.1) can be written in the form

$$\begin{bmatrix} u(n-1) \\ v(n-1) \end{bmatrix} = \begin{bmatrix} 1 & -\alpha(n)e^{in\theta} \\ -\beta(n)e^{-in\theta} & 1 \end{bmatrix} \begin{bmatrix} u(n) \\ v(n) \end{bmatrix}. \tag{1.3.7}$$

Similarly the relations (1.3.2) can be written

$$[1-\alpha(n)\beta(n)] \begin{bmatrix} u(n) \\ v(n) \end{bmatrix} = \begin{bmatrix} 1 & \alpha(n)e^{in\theta} \\ \beta(n)e^{-in\theta} & 1 \end{bmatrix} \begin{bmatrix} u(n-1) \\ v(n-1) \end{bmatrix}. \tag{1.3.7'}$$

This shows that if $D(k) \neq 0$ $k = 0, 1, \ldots, n$ then $u(n)$ and $v(n)$ can be constructed from the $\{\alpha(k)\}_{k=0}^{n}$ and $\{\beta(k)\}_{k=0}^{n}$.

1.4. Szegö's identity

We assume that $D(n)$, $D(n-1) \neq 0$. Let

$$u(n, z) = \sum_{j=0}^{n} \boldsymbol{u}(n, j)z^j, \qquad v(n, z) = \sum_{j=-n}^{0} \boldsymbol{v}(n, j)z^j,$$

and let

$$\frac{u(n, z)}{\boldsymbol{u}(n, 0)} = \prod_{\nu=1}^{n} \left[1 - \frac{z}{z^+(n, \nu)}\right],$$

$$\frac{v(n, z)}{\boldsymbol{v}(n, 0)} = \prod_{\mu=1}^{n} \left[1 - \frac{z^-(n, \mu)}{z}\right].$$

If the degree of $u(n, z)$ is $m < n$ then $n-m$ of the $z^+(n, \nu)$ are ∞, and if the degree of $v(n, z)$ as a polynomial in z^{-1} is $k < n$ then $n-k$ of the $z^-(n, \mu)$ are 0. We define

$$F(n) = \prod_{\mu, \nu=1}^{n} \left[1 - \frac{z^-(n, \mu)}{z^+(n, \nu)}\right].$$

Lemma 4a. If $D(n)$, $D(n-1)$, and $D(n-2) \neq 0$ then

$$F(n) = [1-\alpha(n)\beta(n)]^n F(n-1).$$

If $D(k) \neq 0$ $k = 0, 1, \ldots, n$ then

$$F(n) = \prod_{k=1}^{n} [1-\alpha(k)\beta(k)]^k.$$

Proof. We have

$$F(n) = \prod_{\mu=1}^{n} \frac{u(n, z^-(n, \mu))}{u(n, 0)} = \prod_{\nu=1}^{n} \frac{v(n, z^+(n, \nu))}{v(n, 0)}.$$

By (1.3.1) $u(n, z^-(n, \mu)) = u(n-1, z^-(n, \mu))$ and thus

$$\begin{aligned} F(n) &= \left[\frac{u(n-1, 0)}{u(n, 0)}\right]^n \prod_{\mu=1}^{n} \frac{u(n-1, z^-(n, \mu))}{u(n-1, 0)}, \\ &= \left[\frac{u(n-1, 0)}{u(n, 0)}\right]^n \prod_{\mu=1}^{n} \prod_{\nu=1}^{n-1} \left[1 - \frac{z^-(n, \mu)}{z^+(n-1, \nu)}\right], \\ &= \left[\frac{u(n-1, 0)}{u(n, 0)}\right]^n \prod_{\nu=1}^{n-1} \frac{v(n, z^+(n-1, \nu))}{v(n, 0)}. \end{aligned}$$

By (1.3.2) $v(n, z^+(n-1, \nu)) = [1-\alpha(n)\beta(n)]^{-1} v(n-1, z^+(n-1, \nu))$ so that $F(n)$ is

$$\begin{aligned} &\left[\frac{u(n-1, 0)}{u(n, 0)}\right]^n \left[\frac{v(n-1, 0)}{v(n, 0)}\right]^{n-1} [1-\alpha(n)\beta(n)]^{-n+1} \prod_{\nu=1}^{n-1} \frac{v(n-1, z^+(n-1, \nu))}{v(n-1, 0)} \\ &\qquad = [1-\alpha(n)\beta(n)]^n F(n-1), \end{aligned}$$

as desired. Here we have made use of (1.3.3).

We can now derive (in a somewhat more general form) an identity of Szegö's which is of the greatest importance. Szegö's proof was altogether different, see [66]. The present proof is taken from [35].

Theorem 4b. If $D(n)$, $D(n-1) \neq 0$ then

$$F(n)D(n) = G(n)^{n+1}. \tag{1.4.1}$$

Proof. In the case in which $D(k) \neq 0$ $k = 0, 1, \ldots, n$ this follows from Corollary 3c and Lemma 4a. In order to deal with the general case, in which we assume only that $D(n)$, $D(n-1) \neq 0$, we introduce

$$c(\xi, k) = (1-\xi)\delta(k, 0)+\xi c(k) \qquad k = \ldots, -1, 0, 1, \ldots.$$

Here ξ is a complex parameter. Then $c(o, k) = \delta(k, 0)$ and $c(1, k) = c(k)$. Let

$$D(\xi, n) = \det\,[c(\xi, k-j)]_{k,\, j=0,\ldots,n},$$

etc. It is evident that for $a > 0$ sufficiently small, $|\xi| < a$ implies $D(\xi, k) \neq 0$ $k = 0, 1,\ldots, n$ and therefore $F(\xi, n)D(\xi, n) = G(\xi, n)^{n+1}$. Let $Z(n)$ be the set of zeros of $D(\xi, n)$. $D(\xi, n)$ is a polynomial in ξ and therefore analytic at all finite points. $G(\xi, n)$ is analytic for $\xi \notin Z(n-1)$. We assert that $F(\xi, n)$ is analytic for $\xi \notin Z(n-1) \bigcup Z(n)$. This is because $F(\xi, n)$ is a symmetric function of $\{z^-(\xi, n, \mu)\}_{\mu=0}^n$ and of $\{z^+(\xi, n, \nu)^{-1}\}_{\nu=0}^n$. These are respectively the roots $z^n v(\xi, n, z)$ and $z^n u(\xi, n, z^{-1})$ whose coefficients are rational functions of ξ analytic for $\xi \notin Z(n)$ and whose leading coefficient is different from 0 for $\xi \notin Z(n-1)$. Our desired result now follows by analytic continuation.

Corollary 4c. If $D(n)$, $D(n-1) \neq 0$ then $u(n, z)$ and $z^n v(n, z)$ have no common zeros.

This fact is of cardinal importance.

We derive an alternative expression for $F(n)$ which will be useful later. We assume throughout that $D(n)$, $D(n-1) \neq 0$. Let s be a simple closed rectifiable curve such that 0 and the zeros of $v(n, z)$ lie inside s while the zeros of $u(n, z)$ lie outside s. Then

$$F(n) = \exp\left\{\frac{1}{2\pi i}\int_s \log u(n, z)\,\frac{v(n, z)'}{v(n, z)}\,dz\right\}, \tag{1.4.2}$$

where the integration proceeds in the counterclockwise direction around s. Here $v(n, z)' = (d/dz)v(n, z)$. Since $v(n, z)'/v(n, z)$ has simple poles of residue 1 at $\{z^-(n, \mu)\}_{\mu=1}^n$ plus a simple pole of residue $-n$ at 0 we see that the right hand side of (1.4.2) is equal to

$$\prod_{\mu=1}^{n} \frac{u(n, z^-(n, \mu))}{u(n, 0)} = F(n).$$

See the proof of Lemma 4a.

Let $D(n)$, $D(n-1) \neq 0$. Since $u(n, z)$ and $z^n v(n, z)$ have no common zero the relation

$$p(z)u(n, z) = q(z)z^n v(n, z), \tag{1.4.3}$$

where $p(z)$ and $q(z)$ are polynomials of formal degree $n-1$, is possible only if p and q are identically zero. The relation (1.4.3) is equivalent to a homogeneous system of $2n$ equations with the coefficients of $p(z)$ and $q(z)$ as unknowns. Since this system has only the trivial solution its determinant must be different from 0. We have thus proved the following.

Corollary 4d. If $D(n)$, $D(n-1) \neq 0$ then $\det S(n) \neq 0$ where $S(n)$ is the $2n \times 2n$ matrix

$$S(n) = \begin{bmatrix} \boldsymbol{u}(n,n)\boldsymbol{u}(n,n-1)\ldots \boldsymbol{u}(n,1)\boldsymbol{u}(n,0) & & \\ \qquad \boldsymbol{u}(n,n) \quad \ldots \boldsymbol{u}(n,2)\boldsymbol{u}(n,1)\boldsymbol{u}(n,0) & & \\ \cdot\;\cdot\;\cdot\;\cdot\;\cdot\;\cdot\;\vdots\;\cdot\;\cdot\;\cdot\;\cdot\;\cdot\;\cdot & & \\ \qquad\qquad \boldsymbol{u}(n,n)\boldsymbol{u}(n,n-1)\boldsymbol{u}(n,n-2)\ldots \boldsymbol{u}(n,0) & & \\ \boldsymbol{v}(n,0)\boldsymbol{v}(n,-1) \quad \ldots \boldsymbol{v}(n,-n+1)\boldsymbol{v}(n,-n) & & \\ \qquad \boldsymbol{v}(n,0) \quad \ldots \boldsymbol{v}(n,-n+2)\boldsymbol{v}(n,-n+1)\boldsymbol{v}(n,-n) & & \\ \cdot\;\cdot\;\cdot\;\cdot\;\cdot\;\cdot\;\vdots\;\cdot\;\cdot\;\cdot\;\cdot\;\cdot\;\cdot & & \\ \qquad\qquad \boldsymbol{v}(n,0)\boldsymbol{v}(n,-1)\boldsymbol{v}(n,-2)\ldots \boldsymbol{v}(n,-n) & & \end{bmatrix}$$

Here the entries not indicated are zero.

$\det S(n)$ is the Sylvester resultant of $u(n, z)$ and $z^n v(n, z)$. It can be shown that

$$\det S(n) = \left[\frac{D(n-1)}{D(n)}\right]^{2n} F(n),$$

see [68; p. 91].

1.5. SOME BASIC IDENTITIES

Suppose $D(n)$, $D(n-1) \neq 0$. We know from 1.4 that $u(n, z)$ and $z^n v(n, z)$ have no common zeros. Let, as in 1.4, s be a simple closed rectifiable curve such that 0 and the zeros of $z^n v(n, z)$ are inside s, while the zeros of $u(n, z)$ are outside. Let $f(z)$ be a function analytic in an open set containing s. For any integer m we define

$$E_s^+(m)f \cdot (z) = z^m \frac{1}{2\pi i}\int_s f(\xi)\xi^{-m}\frac{d\xi}{\xi - z} \tag{1.5.1}$$

where z is inside s. Our assumption concerning $f(z)$ and Cauchy's theorem show that s may be deformed to a slightly larger contour, from which it follows that $z^{-m}E_s^+(m)f\cdot(z)$ is analytic in a region containing s and its interior. Similarly we set

$$E_s^-(m)f \cdot (z) = z^{m+1}\frac{1}{2\pi i}\int_s f(\xi)\xi^{-m-1}\frac{d\xi}{z-\xi} \tag{1.5.2}$$

where z is outside s. As above $z^{-m}E_s^-(m)f\cdot(z)$ can be seen to be analytic in a region containing s and its exterior. If $f(z)$ is a Laurent polynomial

$$f(z) = \sum_k f(k)z^k \qquad \text{(finite sum)}, \tag{1.5.3}$$

then a simple computation gives

$$E_s^+(m)f\cdot(z) = \sum_{k \geqq m} f(k)z^k, \qquad E_s^-(m)f\cdot(z) = \sum_{k \leqq m} f(k)z^k. \tag{1.5.4}$$

The operators $E_s^+(n)$ and $E_s^-(n)$ have all the properties in the list (1.1.2) when z is written in place of $e^{i\theta}$. These properties can be proved from the definitions (1.5.1) and (1.5.2); however, a simpler way to verify them is as follows. If $f(z)$ is analytic in a neighborhood of s, then by Runge's theorem $f(z)$ can be uniformly approximated in a smaller neighborhood of s by a sequence $\{f_r\}_1^\infty$ of Laurent polynomials. From the fact that $E_s^+(k)E_s^+(m)f_r = E_s^+(\max[m, k])f_r$, which follows from (1.5.3) and (1.5.4), we see, on passing to the limit, that $E_s^+(k)E_s^+(m)f = E_s^+(\max[m, k])f$, etc.

For $D(n)$, $D(n-1) \neq 0$ we define

$$c(n, z) = \frac{d(n)^2}{u(n, z)v(n, z)}.$$

Theorem 5a. If $D(n)$, $D(n-1) \neq 0$ then

$$c(k) = c(n, k) \qquad -n \leqq k \leqq n$$

where

$$c(n, k) = \frac{1}{2\pi i}\int_s c(n, z)z^{-k-1}\,dz.$$

Proof. We have

$$E_s^+(0)c(n)u(n) = \frac{1}{2\pi i}\int_s \frac{\mathrm{d}(n)^2}{v(n,\,\xi)}\,\frac{\mathrm{d}\xi}{\xi - z}$$

for z inside s. Let z be fixed. The integrand is analytic for ξ on and outside s except at $\xi = \infty$ where there is a pole with residue 1. It follows that $E_s^+(0)c(n)u(n) = 1$. This implies in particular that

$$\sum_{j=0}^{n} \boldsymbol{c}(n,\,k-j)\boldsymbol{u}(n,\,j) = \delta(k,\,0) \qquad k = 0,\,\ldots,\,n.$$

Similarly

$$\sum_{j=-n}^{0} \boldsymbol{c}(n,\,k-j)\boldsymbol{v}(n,\,j) = \delta(k,\,0) \qquad k = -n,\,\ldots,\,0.$$

On the other hand by (1.2.3)

$$\sum_{j=0}^{n} \boldsymbol{c}(k-j)\boldsymbol{u}(n,\,j) = \delta(k,\,0) \qquad k = 0,\,\ldots,\,n,$$

$$\sum_{j=-n}^{0} \boldsymbol{c}(k-j)\boldsymbol{v}(n,\,j) = \delta(k,\,0) \qquad k = -n,\,\ldots,\,0.$$

Thus $\{\boldsymbol{c}(n,\,k)\}_{k=-n}^{n}$ and $\{\boldsymbol{c}(k)\}_{k=-n}^{n}$ satisfy the same system of $2n+2$ linear equations. The corresponding $(2n+2)\times(2n+1)$ matrix is

$$\begin{array}{c} \\ \rightarrow \\ \\ \\ \\ \rightarrow \\ \\ \\ \\ \end{array}
\overset{\displaystyle\downarrow\qquad\qquad\qquad\qquad\qquad\qquad\qquad\qquad\qquad\qquad\qquad}{\begin{bmatrix}
\boldsymbol{u}(n,\,n) & \boldsymbol{u}(n,\,n-1) & \cdots & \boldsymbol{u}(n,\,0) & & & \\
 & \boldsymbol{u}(n,\,n) & \cdots & \boldsymbol{u}(n,\,1) & \boldsymbol{u}(n,\,0) & & \\
 & & & \vdots & & & \\
 & & & \boldsymbol{u}(n,\,n) & \boldsymbol{u}(n,\,n-1) & \cdots & \boldsymbol{u}(n,\,0) \\
\boldsymbol{v}(n,\,1) & \boldsymbol{v}(n,\,1) & \cdots & \boldsymbol{v}(n,\,-n) & & & \\
 & \boldsymbol{v}(n,\,0) & \cdots & \boldsymbol{v}(n,\,-n+1) & & & \\
 & & & \vdots & & & \\
 & & & \boldsymbol{v}(n,\,0) & \boldsymbol{v}(n,\,-1) & \cdots & \boldsymbol{v}(n,\,-n)
\end{bmatrix}}.$$

If we cross out the rows and column indicated by arrows, the remaining $2n\times 2n$ matrix is $S(n)$. That the matrix above is of rank $2n+1$ now follows from Corollary 4d and the fact that $\boldsymbol{v}(n,\,0) \neq 0$. Our proof is a consequence of this fact.

Versions of this result are given in Geronimus [23] and in Baxter [5]. Their demonstrations, however, are different. For still another proof see [34].

Let us now assume that we are given

$$u(z) = \sum_{k=0}^{n} \boldsymbol{u}(k)z^k \qquad \boldsymbol{u}(0) \neq 0,$$

$$v(z) = \sum_{k=-n}^{0} \boldsymbol{v}(k)z^k \qquad \boldsymbol{v}(0) \neq 0,$$

and that $u(z)$ and $z^n v(z)$ have no common zeros. Let s be (as before) a simple closed rectifiable curve such that 0 and the zeros of $v(z)$ lie inside s while the zeros of $u(z)$ lie outside s. We set

$$c(z) = \frac{1}{u(z)v(z)}.$$

The following fundamental identity is essentially due to Shinbrot [60].

Theorem 5b. Under the above assumptions let

$$W_s^+(m)f = E_s^+(0)E_s^-(m)cf,$$
$$Y_s^+(m)f = vE_s^-(m)uv^{-1}E_s^+(0)vf,$$

where $f(z) = \sum_0^m \boldsymbol{f}(k)z^k$. Then if $m \geqq n$

$$W_s^+(m)Y_s^+(m)f = Y_s^+(m)W_s^+(m)f = f.$$

Similarly let

$$W_s^-(m)f = E_s^-(0)E_s^+(-m)cf,$$
$$Y_s^-(m)f = uE_s^+(-m)u^{-1}vE_s^-(0)uf,$$

where $f(z) = \sum_{-m}^0 \boldsymbol{f}(k)z^k$. Then if $m \geqq n$

$$W_s^-(m)Y_s^-(m)f = Y_s^-(m)W_s^-(m)f = f.$$

Proof. We have

$$c(z)f(z) = g_1(z)+g(z)+g_2(z)$$

where $g_1 = E_s^-(-1)cf$, $g = E_s^+(0)E_s^-(m)cf$, $g_2 = E_s^+(m+1)cf$. Since $c = u^{-1}v^{-1}$

$$u^{-1}f = vg_1+vg+vg_2.$$

Applying $E_s^+(0)$ we obtain

$$u^{-1}f = 0+E_s^+(0)vg+vg_2.$$

Multiplying by uv^{-1} this becomes

$$v^{-1}f = uv^{-1}E_s^+(0)vg+ug_2.$$

Applying $vE_s^-(m)$ we find that

$$vE_s^-(m)v^{-1}f = vE_s^-(m)v^{-1}uE_s^+(0)vg.$$

Since

$$vE_s^-(m)v^{-1}f = f - vE_s^+(m+1)v^{-1}f = f - 0$$

we have

$$f = Y_s^+(m)g.$$

Thus $Y_s^+(m)W_s^+(m)f = f$, etc.

Combining Theorems 5a and 5b we obtain the following explicit inversion formula for $W^+(n)$.

Theorem 5c. Let $D(n)$, $D(n-1) \neq 0$. For $f(z) = \sum_0^n f(k)z^k$ let

$$X_s^+(n)f = d(n)^{-2}v(n)E_s^-(n)v(n)^{-1}u(n)E_s^+(0)v(n)f;$$

then

$$W^+(n)X_s^+(n)f = X_s^+(n)W^+(n)f = f.$$

Similarly for $f(z) = \sum_{-n}^0 f(k)z_k$ let

$$X_s^-(n)f = d(n)^{-2}u(n)E_s^+(-n)u(n)^{-1}v(n)E_s^-(0)u(n)f;$$

then

$$W^-(n)X_s^-f = X_s^-(n)W^-(n)f = f.$$

See [10].

2. Finite Toeplitz Operators on Banach Algebras

2.1. Introduction

In this chapter we will, under suitable assumptions, study the behaviour of $W^+(n)$ and $W^-(n)$ as n increases to infinity.

Let A_0 be the Banach space of those complex valued functions $f(k)$ on I for which $||f||_0$ is finite, where

$$||f||_0 = \sum_{k=-\infty}^{\infty} |f(k)|.$$

If we take convolution

$$f * g \cdot (k) = \sum_{j=-\infty}^{\infty} f(k-j) * g(j),$$

as multiplication then A_0 is a Banach algebra.

Definition 1a. A Banach algebra $\boldsymbol{A}$ of complex functions on I with norm $||.||$ is said to be of type $\mathfrak{M}$ if:

i. $\boldsymbol{A}$ is a subalgebra of $\boldsymbol{A}_0$ and $||\boldsymbol{f}|| \geqq ||\boldsymbol{f}||_0$ for every $\boldsymbol{f} \in \boldsymbol{A}$;

ii. every function $\boldsymbol{f}(k)$ with compact support is in $\boldsymbol{A}$ and such functions are dense in $\boldsymbol{A}$;

iii. there exists a constant M such that for every n and all $\boldsymbol{f} \in \boldsymbol{A}$ the following inequalities hold,

$$||\boldsymbol{E}^+(n)\boldsymbol{f}|| \leqq M||\boldsymbol{f}||, \qquad ||\boldsymbol{E}^-(n)\boldsymbol{f}|| \leqq M||\boldsymbol{f}||,$$

where (as in Chapter 1)

$$\boldsymbol{E}^+(n)\boldsymbol{f}\cdot(k) = \begin{cases} \boldsymbol{f}(k) & k \geqq n \\ 0 & k < n \end{cases}, \qquad \boldsymbol{E}^-\boldsymbol{f}\cdot(k) = \begin{cases} \boldsymbol{f}(k) & k \leqq n \\ 0 & k > n \end{cases}.$$

Note that it follows from ii. and iii. that

$$\lim_{n\to\infty} ||\boldsymbol{E}^+(n)\boldsymbol{f}|| = 0, \quad \lim_{n\to-\infty} ||\boldsymbol{E}^-(n)\boldsymbol{f}|| = 0 \qquad \text{for all } \boldsymbol{f} \in \boldsymbol{A}. \tag{2.1.1}$$

Let $v(n)$ be a function defined on I and satisfying the conditions

$$\begin{aligned} v(n) &\geqq 1 && \text{for all } n; \\ v(n+m) &\leqq v(n)v(m) && \text{for all } m, n. \end{aligned}$$

The set of those functions $\boldsymbol{f}(n)$ for which

$$||\boldsymbol{f}|| = \sum_{-\infty}^{\infty} |\boldsymbol{f}(n)|v(n)$$

is finite is a Banach algebra of type $\mathfrak{M}$. In particular $\boldsymbol{A}_0$ is of type $\mathfrak{M}$. These algebras were introduced by Beurling and Gelfand.

Let us denote by $\boldsymbol{A}[a, b]$ the set of elements $\boldsymbol{f}$ in $\boldsymbol{A}$ with support in $[a, b]$. $\boldsymbol{A}[a, b]$ is a Banach space, but not in general a Banach algebra. However, $\boldsymbol{A}[0, \infty]$ and $\boldsymbol{A}[-\infty, 0]$ are Banach algebras. For $\boldsymbol{c}$ in $\boldsymbol{A}$ fixed we define

$$\begin{aligned} \boldsymbol{W}^+(n)\boldsymbol{f} &= \boldsymbol{E}^+(0)\boldsymbol{E}^-(n)\boldsymbol{c} * \boldsymbol{f} && \boldsymbol{f} \in \boldsymbol{A}[0, n], \\ \boldsymbol{W}^-(n)\boldsymbol{f} &= \boldsymbol{E}^-(0)\boldsymbol{E}^+(-n)\boldsymbol{c} * \boldsymbol{f} && \boldsymbol{f} \in \boldsymbol{A}[-n, 0]. \end{aligned}$$

It is evident from our assumptions that

$$||\boldsymbol{W}^+(n)|| \leqq M^2||\boldsymbol{c}||, \qquad ||\boldsymbol{W}^-(n)|| \leqq M^2||\boldsymbol{c}||,$$

where $||\boldsymbol{W}^+(n)||$ is the norm of $\boldsymbol{W}^+(n)$ as a linear transformation of the

Banach space $A[0, n]$ into itself, etc. We further define the half-infinite Toeplitz operators (also called Wiener-Hopf operators)

$$W^+(\infty)f = E^+(0)c * f \qquad f \in A[0, \infty],$$

and

$$W^-(\infty)f = E^-(0)c * f \qquad f \in A[-\infty, 0].$$

It is easy to see that $||W^+(\infty)||, ||W^-(\infty)|| \leqq M||c||$.

For A of type $\mathfrak{M}$ we define $\mathscr{A}$ to be the set of all functions $f(e^{i\theta})$ on T of the form

$$f(e^{i\theta}) = \sum_{-\infty}^{\infty} f(k)e^{ik\theta} = \mathscr{F}f.$$

If we set $||f|| = ||f||$ then $\mathscr{A}$ is a Banach algebra of functions on T. The inverse mapping $\mathscr{F}^{-1}$ is given by

$$\mathscr{F}^{-1}f \cdot (k) = \frac{1}{2\pi}\int_T f(e^{i\theta})e^{-ik\theta}d\theta = f(k).$$

It is evident that $\mathscr{F}^{-1}$ maps $\mathscr{A}$ isometrically onto A. $\mathscr{A}$ of type $\mathfrak{M}$ means that A is of type $\mathfrak{M}$. The various operators $E^+(n), E^-(n)$, $W^+(n)$, etc. are transferred to $\mathscr{A}$ in the usual way; for example

$$W^+(n)f = \mathscr{F}W^+(n)\mathscr{F}^{-1}f \qquad f \in \mathscr{A}[0, n].$$

Here $\mathscr{A}[0, n]$ is the image of $A[0, n]$ under $\mathscr{F}$, etc.

In (2.2.) we will show that if $\mathscr{A} \in \mathfrak{M}$ and if $W^+(\infty)$ and $W^-(\infty)$ are invertible then $W^+(n)$ and $W^-(n)$ are uniformly invertible for all sufficiently large n; that is, there exists an integer n_0 and a positive constant A such that if $n \geqq n_0$ then $W^+(n)^{-1}$ and $W^-(n)^{-1}$ exist and

$$||W^+(n)^{-1}|| \leqq A, \qquad ||W^-(n)^{-1}|| \leqq A. \tag{2.1.2}$$

These are the 'finite section' inequalities of Baxter [8]. Using these Baxter obtained a very complete and elegant theory of finite Toeplitz operators and Szegö polynomials for the Beurling Gelfand algebras of type $\mathfrak{M}$, see [5] and [6]. In sections 2–7 we will carry out Baxter's program for a somewhat more general class of Banach algebras and with many additions and changes.

2.2. The inversion of half-infinite Toeplitz operators

In order to prove the finite section inequalities we must first analyze the conditions under which $W^+(\infty)$ and $W^-(\infty)$ are invertible.

Theorem 2a. Let $\mathscr{A}$ be of type $\mathfrak{M}$. Then $W^+(\infty)$ and $W^-(\infty)$ are both invertible if and only if

$$c = \mathrm{d}^2 U^{-1} V^{-1}$$

where $d \neq 0$ and where

$$U,\ U^{-1} \in \mathscr{A}[0, \infty], \qquad V,\ V^{-1} \in \mathscr{A}[-\infty, 0].$$

Proof. Necessity. Note that we use only the fact that $W^+(\infty)$ and $W^-(\infty)$ have 1 in their range. By assumption there exist functions $U \in \mathscr{A}[0, \infty]$ and $V \in \mathscr{A}[-\infty, 0]$ such that

$$W^+(\infty)U = 1, \qquad W^-(\infty)V = 1. \tag{2.2.1}$$

We have

$$U(\mathrm{e}^{i\theta}) = \sum_{k \geqq 0} U(k)\mathrm{e}^{ik\theta}, \qquad V(\mathrm{e}^{i\theta}) = \sum_{k \leqq 0} V(k)\mathrm{e}^{ik\theta}.$$

Let

$$V_1(\mathrm{e}^{i\theta}) = c(\mathrm{e}^{i\theta})U(\mathrm{e}^{i\theta}), \qquad U_1(\mathrm{e}^{i\theta}) = c(\mathrm{e}^{i\theta})V(\mathrm{e}^{i\theta}). \tag{2.2.2}$$

Then it follows from (2.2.1) that

$$V_1(\mathrm{e}^{i\theta}) = \sum_{k \leqq 0} V_1(k)\mathrm{e}^{ik\theta} \qquad V_1(0) = 1,$$
$$U_1(\mathrm{e}^{i\theta}) = \sum_{k \geqq 0} U_1(k)\mathrm{e}^{ik\theta} \qquad U_1(0) = 1.$$

From (2.2.2) we see that

$$VV_1 = cUV = UU_1.$$

The (absolutely convergent) Fourier series for VV_1 involves only the exponentials $\{e^{ik\theta}\}_{k \leqslant 0}$ and the Fourier series for UU_1 involves only the exponentials $\{e^{ik\theta}\}_{k \geqslant 0}$. It follows that there exists a constant d^2 such that

$$V_1(\mathrm{e}^{i\theta})V(\mathrm{e}^{i\theta}) = d^2 = U_1(\mathrm{e}^{i\theta})U(\mathrm{e}^{i\theta}).$$

From (2.2.2) we have $cUU_1^* = V_1 U_1^*$. Since

$$\frac{1}{2\pi}\int_T V_1 U_1^* \mathrm{d}\theta = 1$$

it follows that $V_1(e^{i\theta})U_1(e^{i\theta})^* \not\equiv 0$ and hence that $U(e^{i\theta})U_1(e^{i\theta})^* \not\equiv 0$. Consequently $U(e^{i\theta})U_1(e^{i\theta}) \not\equiv 0$ and $\mathrm{d}^2 \neq 0$. Since $U(e^{i\theta})^{-1} = \mathrm{d}^{-2}U_1(e^{i\theta})$ we have $U^{-1} \in A[0, \infty]$. Similarly $V^{-1} \in A[-\infty, 0]$.

Sufficiency. We define operators $X^+(\infty)$ and $X^-(\infty)$ on $\mathscr{A}[0, \infty]$ and $\mathscr{A}[-\infty, 0]$ respectively by the formulas

$$X^+(\infty)f = \mathrm{d}^{-2}\,UE^+(0)Vf \qquad f \in \mathscr{A}[0, \infty],$$
$$X^-(\infty)f = \mathrm{d}^{-2}\,VE^-(0)Uf \qquad f \in \mathscr{A}[-\infty, 0].$$

We assert that $X^+(\infty)$ is the inverse of $W^+(\infty)$ and $X^-(\infty)$ is the inverse of $W^-(\infty)$. Suppose that

$$W^+(\infty)f = g \qquad f \in \mathscr{A}[0, \infty].$$

Then

$$cf = g_1 + g$$

where $g_1 \in \mathscr{A}[-\infty, -1]$. It follows that

$$\mathrm{d}^2\, U^{-1}f = Vg_1 + Vg.$$

Applying $E^+(0)$ we obtain

$$\mathrm{d}^2\, U^{-1}f = E^+(0)Vg.$$

Thus $X^+(\infty)W^+(\infty)f = f$ for $f \in \mathscr{A}[0, \infty]$. A similar computation shows that $W^+(\infty)X^+(\infty)f = f$ for $f \in \mathscr{A}[0, \infty]$, etc. Compare in these connections [12].

We conclude this section by giving some convenient necessary and sufficient conditions on c for $W^+(\infty)^{-1}$ and $W^-(\infty)^{-1}$ to exist.

Theorem 2b. If $\mathscr{A} \in \mathfrak{M}$ then the maximal ideal space $\mathscr{M}(\mathscr{A})$ of $\mathscr{A}$ can be identified with an annulus $\{z | \rho_1 \leqq |z| \leqq \rho_2\}$ in the complex z-plane where $0 < \rho_1 \leqq 1 \leqq \rho_2 < \infty$. The multiplicative linear functional λ_z corresponding to z in $\mathscr{M}(\mathscr{A})$ is given by

$$\lambda_z(f) = \sum_{-\infty}^{\infty} f(n)z^n = f(z).$$

For the (routine) demonstration of this result see [29].

Let us denote by $I(\mathscr{A})$ the set of $c \in \mathscr{A}$ for which $W^+(\infty)$ and $W^-(\infty)$ are both invertible.

Theorem 2c. If $\mathscr{A} \in \mathfrak{M}$ then $c \in I(\mathscr{A})$ if and only if $\lambda_z(c) \neq 0$ for $z \in \mathscr{M}(\mathscr{A})$ and $[\arg c(e^{i\theta})]_{-\pi}^{\pi} = 0$.

Proof. The necessity of these conditions follows from Theorem 2a.

Since $\log c(z)$ is analytic for $\rho_1 < |z| < \rho_2$ and continuous for $\rho_1 \leqq |z| \leqq \rho_2$ we can, by Runge's theorem, find a Laurent polynomial. $p(z) = \sum_{-r}^{r} p(k)z^k$ such that $|p(z)+\log c(z)| < \pi/2$ for $\rho_1 \leqq |z| \leqq \rho_2$. This implies that $Rl\,[c(z) \exp p(z)] > 0$ for $\rho_1 \leqq |z| \leqq \rho_2$. The Wiener Lévy theorem may therefore be applied to show that

$$\log c = \log [ce^p] - p$$

is in $\mathscr{A}$. It is now apparent that if

$$U = \exp[-E^+(0) \log c], \qquad V = \exp[-E^-(0) \log c],$$
$$\mathrm{d}^2 = \exp[-E^-(0)E^+(0) \log c],$$

then

$$U, U^{-1} \in E^+(0)\mathscr{A}, \qquad V, V^{-1} \in E^-(0)\mathscr{A}, \qquad \mathrm{d} \neq 0,$$

and

$$c = \mathrm{d}^2 U^{-1} V^{-1}.$$

Alternatively we could use the Arens-Calderon form of the Wiener-Lévy theorem [12], which is valid (under suitable assumptions) on Riemann surfaces.

2.3. The finite section inequalities

We will make repeated use of the following well-known result. See Hille and Phillips [25; p. 118].

Theorem 3a. Let B be a (possibly non-commutative) Banach algebra with an identity. If $b_0 \in B$ is invertible then b is invertible if

$$||b - b_0|| \leqq ||b_0^{-1}||^{-1},$$

and we have

$$||b^{-1} - b_0^{-1}|| \leqq \frac{||b_0^{-1}||^2 ||b - b_0||}{1 - ||b - b_0|| \, ||b_0^{-1}||}.$$

All Banach algebras $\mathscr{A}$ in this section will be of type $\mathfrak{M}$.

Theorem 3b. Let $c \in I(\mathscr{A})$. Then there exists an integer n_0 and a constant A such that $W^+(n)^{-1}$ and $W^-(n)^{-1}$ are defined for $n \geqq n_0$ and

$$||W^+(n)^{-1}|| \leqq A, \; ||W^-(n)^{-1}|| \leqq A \qquad \text{for} \quad n \geqq n_0.$$

Proof. By Theorem 2a

$$c = \mathrm{d}^2/UV$$

where $d^2 \neq 0$, U, $U^{-1} \in \mathscr{A}[0, \infty]$ and $V, V^{-1} \in \mathscr{A}[-\infty, 0]$. Let

$$U_m(\mathrm{e}^{i\theta}) = \sum_{k=0}^{m} U(k)\mathrm{e}^{ik\theta}, \qquad V_m(\mathrm{e}^{i\theta}) = \sum_{k=-m}^{0} V(k)\mathrm{e}^{ik\theta}.$$

Clearly

$$\lim_{m\to\infty} ||U - U_m|| = \lim_{m\to\infty} ||V - V_m|| = 0.$$

Since both $\mathscr{A}[0, \infty]$ and $\mathscr{A}[-\infty, 0]$ are Banach algebras with an identity it follows from Theorem 3a that there exists an integer m_0 such that $U_m^{-1} \in \mathscr{A}[0, \infty]$ and $V_m^{-1} \in \mathscr{A}[-\infty, 0]$ for all $m \geqq m_0$. If we define

$$c_m(\mathrm{e}^{i\theta}) = d^2/U_m(\mathrm{e}^{i\theta})V_m(\mathrm{e}^{i\theta}),$$

$$W_m^+(n)f = E^+(0)E^-(n)c_m f \qquad \text{for} \quad f \in \mathscr{A}[0, n],$$

then by Shinbrot's identity, Theorem 5b of Chapter I, with s the unit circle, $W_m^+(n)$ is inverted by the operator $Y_m^+(n)$ for $n \geqq m \geqq m_0$ where

$$Y_m^+(n)f = d^{-2}V_m E^-(n)U_m V_m^{-1}E^+(0)V_m f.$$

For $n \geqq m \geqq m_0$ we have

$$||Y_m^+(n)|| \leqq d^{-2}||V_m||^2||V_m^{-1}||\,||U_m|| = D_1(m),$$

where $\overline{\lim}_{m\to\infty} D_1(m) < \infty$. On the other hand for $m \geqq m_0$ and all n

$$||W^+(n) - W_m^+(n)|| \leqq M^2 \left\| \frac{d^2}{UV} - \frac{d^2}{U_m V_m} \right\| = D_2(m)$$

where $\lim_{m\to\infty} D_2(m) = 0$. It follows from Theorem 3a applied to the non-commutative Banach algebra of all linear transformations of $\mathscr{A}[0, n]$ into itself taken with the operator norm, that if $n_0 \geqq m_0$ is such that

$$D_1(n_0)D_2(n_0) < 1,$$

then $W^+(n)$ is uniformly invertible for all $n \geqq n_0$. A similar argument applies to $W^-(n)$.

The proof of Theorem 3b given here is taken from [33]. Baxter's original demonstration [8] is very elegant and definitely simpler;

however, the proof given above can be adapted more readily to other situations.

Corollary 3c. Let $c \in I(\mathscr{A})$, let $g \in \mathscr{A}[0, \infty]$, and let n_0 be as in Theorem 3b. If

$$f = W^+(\infty)^{-1}g, \qquad f_n = W^+(n)^{-1}E^-(n)g \qquad n \geqq n_0$$

then there is a constant A independent of g such that

$$||f - f_n|| \leqq A||E^+(n+1)f|| \qquad n \geqq n_0.$$

Proof. It is easy to see that

$$W^+(n)[f_n - E^-(n)f] = E^+(0)E^-(n)[cE^+(n+1)f]$$

from which it follows (Theorem 3b) that $||f_n - E^-(n)f|| \leqq A(||E^+(n+1)f||)$, etc.

Corollary 3d. If $c \in I(\mathscr{A})$ then

$$\lim_{n\to\infty} ||U(e^{i\theta}) - u(n, e^{i\theta})|| = 0,$$

$$\lim_{n\to\infty} ||V(e^{i\theta}) - v(n, e^{i\theta})|| = 0.$$

Proof. This is left to the reader.

Corollary 3e. If $c \in I(\mathscr{A}_0)$ then

$$\lim_{n\to\infty} D(n)/D(n-1) = \exp\left[\frac{1}{2\pi}\int_T \log c(e^{i\theta})d\theta\right].$$

Proof. By Corollary 3d $\lim_{n\to\infty} \boldsymbol{u}(n, 0) = \boldsymbol{U}(0) = d^2$ and thus since $\boldsymbol{u}(n, 0) = D(n-1)/D(n)$

$$\lim_{n\to\infty} D(n)/D(n-1) = \lim_{n\to\infty} d(n)^{-2} = d^{-2}.$$

The function $U \in \mathscr{A}_0[0, \infty]$. Since $U^{-1} \in \mathscr{A}_0[0, \infty]$ it follows that $U(z) \neq 0$ for $|z| \leqq 1$. Consequently the Poisson-Jensen formula, [67, p. 124], gives

$$\frac{1}{2\pi}\int_T \log U(e^{i\theta})d\theta = \log \boldsymbol{U}(0) = 2 \log d,$$

and a similar argument shows that

$$\frac{1}{2\pi}\int_T \log V(e^{i\theta})d\theta = \log V(0) = 2\log d.$$

We have

$$\frac{1}{2\pi}\int_T \log c(e^{i\theta})d\theta = \frac{1}{2\pi}\int_T [2\log d - \log U(e^{i\theta}) - \log V(e^{i\theta})]d\theta,$$
$$= -2\log d,$$

and thus

$$\exp\left[\frac{1}{2\pi}\int_T \log c(e^{i\theta})d\theta\right] = d^{-2}.$$

See Baxter [9].

Let $c(d\theta)$ be a finite non-negative measure on the Borel sets of T, and let

$$c(k) = \frac{1}{2\pi}\int_T e^{-ik\theta}c(d\theta) \qquad k = \ldots, -1, 0, 1, \ldots,$$
$$D(n) = \det\,[c(k-j)]_{k,\, j=0,\ldots,n}.$$

Then it can be shown that

$$\lim_{n\to\infty} D(n)/D(n-1) = \exp\left\{\frac{1}{2\pi}\int_T \log c_a(e^{i\theta})d\theta\right\},$$

where

$$c(d\theta) = c_a(e^{i\theta})d\theta + c_s(d\theta)$$

is the decomposition of $c(d\theta)$ into its absolutely continuous and singular parts. This result was first established by Szegö in the case $c_s \equiv 0$; see [64]. The general case is due to Kolmogoroff [45] and Krein [46]. This relation is called the Szegö limit theorem. Corollary 3e, which is due to Baxter, was the first *complex* Szegö limit theorem. See section 2.10.

2.4. The '#' form of the finite section inequalities

In this section we will consider a more restricted class of Banach algebras and obtain for them a stronger version of the finite section inequalities.

For $\boldsymbol{f}, \boldsymbol{g} \in A_0$ we write

$$\boldsymbol{f} \prec \boldsymbol{g}$$

if

$$|f(k)| \leqq |g(k)| \qquad k = \ldots, -1, 0, 1, \ldots.$$

We also set

$$f^{\#}(k) = |f(k)| \qquad k = \ldots, -1, 0, 1, \ldots.$$

The following properties are an immediate consequence of our definitions:

$$f_1 \prec g_1 \quad \text{and} \quad f_2 \prec g_2 \Rightarrow \begin{matrix} f_1 + f_2 \prec g_1^{\#} + g_2^{\#} \\ f_1 * f_2 \prec g_1^{\#} * g_2^{\#} \end{matrix}.$$

Definition 4a. A Banach algebra A of complex functions on I with norm $||.||$ is said to be type $\mathfrak{S}$ if:

i. A is a subalgebra of A_0 and $||f|| \geqq ||f||_0$ for every $f \in A$;

ii. every function $f(k)$ with compact support is in A and such functions are dense in A;

iii. $f \in A$, $g \in A_0$, and $g \prec f$ implies that $g \in A$ and $||g|| \leqq ||f||$.

It follows from iii. that

$$||E^{+}(n)f|| \leqq ||f||, \; ||E^{-}(n)f|| \leqq ||f|| \qquad \text{for all} \quad f \in A.$$

Consequently every Banach algebra of type $\mathfrak{S}$ is of type $\mathfrak{M}$. Let $\mathscr{A}$ be defined as in section 2.1. Then $f \prec g$ means that $f \prec g$, and $f^{\#} = \mathscr{F}[(\mathscr{F}^{-1}f)^{\#}]$. $\mathscr{A}$ will be said to be of type $\mathfrak{S}$ if A is of type $\mathfrak{S}$.

The Beurling-Gelfand algebras of section 2.1 are all of type $\mathfrak{S}$. Another example, to which we will return later, is as follows. Let

$$||f|| = 3\{\sum_{-\infty}^{\infty} |f(k)| + [\sum_{-\infty}^{\infty} |f(k)^2 k|]^{\frac{1}{2}}\}; \tag{2.4.1}$$

then the set of all f for which $||f||$ is finite is a Banach algebra of type $\mathfrak{S}$.

Lemma 4b. Let $\mathscr{A}$ be of type $\mathfrak{S}$ and let $f, a, b \in \mathscr{A}$. If

i. $f \prec a^{\#}f^{\#} + b^{\#}$,

ii. $||a^{\#}|| < 1$,

then

$$f \prec \{\sum_{j=0}^{\infty} (a^{\#})^j\} b^{\#}.$$

Proof. Iterating *i.* we find successively that

$$f \prec (a^{\#})^2 f^{\#} + a^{\#} b^{\#} + b^{\#},$$
$$f \prec (a^{\#})^3 f^{\#} + (a^{\#})^2 b^{\#} + a^{\#} b^{\#} + b^{\#},$$
$$\vdots$$
$$f \prec (a^{\#})^n f^{\#} + \sum_{j=0}^{n-1} (a^{\#})^j b .$$

If we now let $n \to \infty$ we obtain our desired result.

Theorem 4c. Let $c \in I(\mathscr{A})$ where $\mathscr{A} \in \mathfrak{S}$. Then there exists an integer n_1 and a function $C = C^{\#}$ in $\mathscr{A}$ such that if $n \geqq n_1$

$$f \prec C[W^+(n)f]^{\#} \qquad \text{for all} \quad f \in \mathscr{A}[0, n],$$
$$f \prec C[W^-(n)f]^{\#} \qquad \text{for all} \quad f \in \mathscr{A}[-n, 0].$$

Proof. Let $W_m^+(n)$, $Y_m^+(n)$, and m_0 be as in the proof of Theorem 3b. Let $f \in \mathscr{A}[0, n]$ and let $g = W^+(n)f$. If $n \geqq m \geqq m_0$ then since

$$W_m^+(n)f = [W_m^+(n) - W^+(n)]f + g,$$

we have

$$f = Y_m^+(n)[W_m^+(n) - W^+(n)]f + Y_m^+(n)g,$$

which implies that

$$f \prec a^{\#} f^{\#} + b ,$$

where

$$a^{\#} = |d|^{-2}(V_m^{\#})^2(V_m^{-1})^{\#} U_m^{\#}[c - c_m]^{\#},$$
$$b^{\#} = |d|^{-2}(V_m^{\#})^2(V_m^{-1})^{\#} U_m g^{\#}.$$

We have

$$\|a^{\#}\| \leqq |d|^{-2} \|V_m\|^2 \|V_m^{-1}\| \, \|U_m\| \, \|c - c_m\| = D_3(m)$$

where $\lim_{m \to \infty} D_3(m) = 0$. Choose $n_1 \geqq m_0$ so large that $D_3(n_1) < 1$, and apply Lemma 4b.

This result is due to the author [31].

2.5. Expansions in Szegö polynomials, I

Let $c \in \mathscr{A}$ where $\mathscr{A} \in \mathfrak{M}$. If $D(k), D(k-1) \neq 0$, so that $\varphi(k, e^{i\theta})$ and $\psi(k, e^{i\theta})$ are defined, the formal biorthogonality of Theorem 2a, Chapter I, becomes a genuine biorthogonality and we have

$$\begin{aligned} &\frac{1}{2\pi}\int_T \varphi(k, e^{i\theta})e^{-ij\theta}c(e^{i\theta})d\theta = 0, \quad j = 0, 1, \ldots, k-1, \\ &\frac{1}{2\pi}\int_T \psi(k, e^{i\theta})e^{ij\theta}c(e^{i\theta})d\theta = 0, \quad j = 0, 1, \ldots, k-1, \end{aligned} \tag{2.5.1}$$

and

$$\frac{1}{2\pi}\int_T \varphi(k, e^{i\theta})\psi(k, e^{i\theta})c(e^{i\theta})d\theta = 1. \tag{2.5.2}$$

Lemma 5a. Let $\mathscr{A} \in \mathfrak{S}$ and $c \in I(\mathscr{A})$, and let $D(k) \neq 0$ for all $k \geqq N-1$. If for $f \in \mathscr{A}[0, \infty]$ we set

$$\boldsymbol{f}^{\sim}(k) = \frac{1}{2\pi}\int_T f(e^{i\theta})\psi(k, e^{i\theta})c(e^{i\theta})d\theta \qquad k \geqq N,$$

then $\boldsymbol{f}^{\sim} \in A[N, \infty]$ and there exists a constant A independent of f such that

$$\|\boldsymbol{f}^{\sim}\| \leqq A\|f\|.$$

Proof. Let n_1 and C be as in Theorem 4c and let $k \geqq n_1$ (n_1 is necessarily greater than or equal to N). By Theorem 4c we have $u(n, e^{i\theta}) \prec C(e^{i\theta})$ and $v(n, e^{i\theta}) \prec C(e^{i\theta})$ for $n \geqq n_1$. Replacing C by a function with suitably larger coefficients (which we will also denote by C) we can so arrange things that

$$u(n, e^{i\theta}) \prec C(e^{i\theta}), \quad v(n, e^{i\theta}) \prec C(e^{i\theta}) \qquad n \geqq N.$$

It follows from Corollary 3e that there is a constant D such that

$$|d(n)^{-1}| \leqq D \qquad n \geqq N.$$

We have, if $k \geqq N$,

$$f^{\sim}(k) = \frac{1}{2\pi}\int_T f(e^{i\theta})d(k)^{-1}u(k, e^{i\theta})e^{-ik\theta}c(e^{i\theta})d\theta,$$

and thus

$$\begin{aligned} |\boldsymbol{f}^{\sim}(k)| &\leqq \frac{1}{2\pi}\int_T Df^{\#}(e^{i\theta})C(e^{i\theta})c^{\#}(e^{i\theta})e^{-ik\theta}d\theta, \\ &\leqq \boldsymbol{F}(k) \end{aligned}$$

if $F(e^{i\theta}) = DC(e^{i\theta})c^{\#}(e^{i\theta})f^{\#}(e^{i\theta})$, from which our desired result follows with $A = D\|C\|\,\|c\|$.

Let B be a Banach space and $\sum_N^\infty b_k$ a series with terms from B. Let $\mathscr{P}_N$ be the family of all finite subsets of $\{k|N \leqq k < \infty\}$ ordered by inclusion. Our series is said to converge unconditionally if given $\varepsilon > 0$ there exists $p_\epsilon \in \mathscr{P}_N$ such that $p_1, p_2 \geqq p_\epsilon$ implies

$$\|\sum_{k\in p_1} b_k - \sum_{k\in p_2} b_k\| < \varepsilon,$$

etc.

Lemma 5b. Under the assumptions of Lemma 5a the series

$$\sum_{k=N}^{\infty} f^{\sim}(k)\varphi(k, e^{i\theta})$$

converges unconditionally to a function in $\mathscr{A}[0, \infty]$, and there is a constant A independent of f such that

$$\|\sum_{k=N}^{\infty} f^{\sim}(k)\varphi(k, e^{i\theta})\| \leqq A\|f\|.$$

Proof. From

$$\varphi(k, e^{i\theta}) = d(k)^{-1}v(k, e^{i\theta})e^{ik\theta}$$

we see that

$$\varphi(k, e^{i\theta}) \prec DC(e^{i\theta})e^{ik\theta}.$$

Consequently if $p(n) = \{N, N+1, \ldots, n\}$ and if $p_1, p_2 \geqq p(n)$

$$\sum_{p_1} f^{\sim}(k)\varphi(k) - \sum_{p_2} f^{\sim}(k)\varphi(k) \prec DC(e^{i\theta}) \sum_{k=n+1}^{\infty} |f^{\sim}(k)|e^{ik\theta},$$

and thus

$$\|\sum_{p_1} - \sum_{p_2}\| \leqq D\|C\|\,\|E^+(n+1)f^{\sim}\|.$$

Thus our series is unconditionally convergent. Since $|f^{\sim}(k)| \leqq F(k)$ we have

$$\sum_{p_1} f^{\sim}(k)\varphi(k) \prec D^2C(e^{i\theta})^2 c^{\#}(e^{i\theta})f^{\#}(e^{i\theta})$$

It follows that

$$\|\sum_{N}^{\infty} f^{\sim}(k)\varphi(k, e^{i\theta})\| \leqq A\|f\|,$$

where $A = D^2\|C\|^2\|c^{\#}\|$.

Theorem 5c. Suppose $c \in I(\mathscr{A})$ where $\mathscr{A} \in \mathfrak{S}$ and that $D(k) \neq 0$ for $k \geqq N-1$. If $f \in E^+(0)\mathscr{A}$ then the series

$$\sum_{k=N}^{\infty} \boldsymbol{f}^{\sim}(k)\varphi(k, \mathrm{e}^{i\theta}),$$

where

$$\boldsymbol{f}^{\sim}(k) = \frac{1}{2\pi}\int_T f(\mathrm{e}^{i\theta})\psi(k, \mathrm{e}^{i\theta})c(\mathrm{e}^{i\theta})\mathrm{d}\theta,$$

converges unconditionally in $\mathscr{A}$ and

$$E^+(N)[f(\mathrm{e}^{i\theta}) - \sum_{k=N}^{\infty} \boldsymbol{f}^{\sim}(k)\varphi(k, \mathrm{e}^{i\theta})] = 0. \tag{2.5.3}$$

Proof. If, for $f(e^{i\theta}) \in \mathscr{A}[N, \infty]$, we set

$$J_N f \cdot (\mathrm{e}^{i\theta}) = E^+(N) \sum_{k=N}^{\infty} \boldsymbol{f}^{\sim}(k)\varphi(k, \mathrm{e}^{i\theta}),$$

then it follows from Lemma 5b that J_N is a bounded linear transformation of $\mathscr{A}[N, \infty]$ into itself.

If $f(e^{i\theta})$ is a trigonometric polynomial in $\mathscr{A}[N, \infty]$ then, since $\boldsymbol{\varphi}(k, k) \neq 0$ for $k \geqq N$, then exist constants $\boldsymbol{a}(k)$ $k = N, N+1, \ldots, n$ such that

$$E^+(N)[f(\mathrm{e}^{i\theta}) - \sum_{k=N}^{n} \boldsymbol{a}(k)\varphi(k, \mathrm{e}^{i\theta})] = 0,$$

and consequently $J_N f(e^{i\theta}) = f(e^{i\theta})$. Since J_N coincides with I_N, the identity transformation on $\mathscr{A}[N, \infty]$, for all trigonometric polynomials in $\mathscr{A}[N, \infty]$, and since these are a dense set, it follows that $J_N = I_N$, which is what we wished to show.

There is a parallel result concerned with the expansion of a function $f \in \mathscr{A}[-\infty, 0]$ in a series of ψ's. We regard Theorem 5c as including both these cases.

An algebra A is said to be of type $\mathfrak{G}$ if A is a Beurling-Gelfand algebra. See § 1 of II. Every algebra of type $\mathfrak{G}$ is *à fortiori* of type $\mathfrak{S}$ and for algebras of this more restricted type we have the following result.

Theorem 5d. Suppose $c \in I(\mathscr{A})$ where $\mathscr{A} \in \mathfrak{G}$ and that $D(k) \neq 0$ for $k \geqq N-1$. If $f \in E^+(0)\mathscr{A}$ then the series

$$\sum_{k=N}^{\infty} \boldsymbol{f}^{\sim}(k)\varphi(k, \mathrm{e}^{i\theta})$$

converges absolutely in A and

$$E^+(N)[f(e^{i\theta})-\sum_{k=N}^{\infty} f^\sim(k)\varphi(k, e^{i\theta})] = 0.$$

Proof. For $g \in \mathscr{A}$ we have $\|g\| = \sum_{-\infty}^{\infty}|g(k)|v(k)$. Lemma 5a implies that $\sum_N^\infty|f^\sim(k)|v(k) < \infty$, and it follows from the proof of Lemma 5b that $\|\varphi(k, e^{i\theta})\| = 0(v(k))$. Thus

$$\sum_{k=N}^{\infty}|f^\sim(k)|\, \|\varphi(k, e^{i\theta})\| < \infty,$$

etc.

2.6. Expansions in Szegö polynomials, II

We now turn to the case when $\mathscr{A}$ is of type $\mathfrak{M}$. We set

$$S^+(n)f = d(n)^{-2}\{u(n)E^+(0)cv(n)f-v(n)E^+(n+1)cu(n)f\}.$$

Theorem 6a. Let $\mathscr{A}$ be of type $\mathfrak{M}$ and let $D(n)$, $D(n-1) \neq 0$. If $u(n)^{-1} \in \mathscr{A}[0, \infty]$, $v(n)^{-1} \in \mathscr{A}[-\infty, 0]$, then

$$S^+(n)e^{ik\theta} = e^{ik\theta} \qquad 0 \leqq k \leqq n,$$
$$S^+(n)\varphi(k, e^{i\theta}) = 0 \qquad k > n.$$

In the second relation we have implicitly assumed that $D(k)$, $D(k-1) \neq 0$ so that $\varphi(k)$ is defined.

Proof. We recall that

$$cv(n) = v_1(n)+1+v_2(n),$$

where $v_1(n) \in \mathscr{A}[-\infty, -n-1]$, $v_2(n) \in \mathscr{A}[1, \infty]$, and that

$$cu(n) = u_1(n)+1+u_2(n),$$

where $u_1(n) \in \mathscr{A}[-\infty, -1]$, $u_2(n) \in \mathscr{A}[n+1, \infty]$. Let $e(k) = e^{ik\theta}$. If $0 \leqq k \leqq n$ then

$$\begin{aligned} u(n)E^+(0)cv(n)e(k) &= u(n)e(k)E^+(0)cv(n), \\ &= e(k)E^+(0)cu(n)v(n), \end{aligned}$$

and

$$\begin{aligned} v(n)E^+(n+1)cu(n)e(k) &= v(n)e(k)E^+(n+1)cu(n), \\ &= e(k)v(n)e(k)E^+(0)[cu(n)-1], \\ &= e(k)E^+(0)[cu(n)v(n)-d(n)^2], \\ &= e(k)E^+(0)cu(n)v(n)-d(n)^2 e(k). \end{aligned}$$

Combining these results we see that $S^+(n)e(k) = e(k)$ as desired.

We further recall that

$$c\varphi(k) = w_1 + w_2$$

where $w_1 \in \mathscr{A}[-\infty, -1]$, $w_2 \in \mathscr{A}[k, \infty]$. If $k > n$ then

$$u(n)E^+(0)cv(n)\varphi(k) = u(n)v(n)E^+(0)[c\varphi(k)],$$

and

$$v(n)E^+(n+1)cu(n)\varphi(k) = v(n)u(n)E^+(0)c\varphi(k).$$

Together these formulas imply that $S^+(n)\varphi(k) = 0$.

Theorem 6a is closely related to the Christoffel-Darboux formula. See [29].

Theorem 6b. Let $c \in I(\mathscr{A})$ where $\mathscr{A}$ is of type $\mathfrak{M}$ and suppose that $D(k) \neq 0$ for $k \geqq N-1$. If $f \in E^+(0)\mathscr{A}$ then the series

$$\sum_{k \geqq N} f^{\sim}(k)\varphi(k, e^{i\theta}),$$

where

$$f^{\sim}(k) = \frac{1}{2\pi}\int_T f(e^{i\theta})\psi(k, e^{i\theta})c(e^{i\theta})d\theta,$$

converges (conditionally) in $\mathscr{A}$, and

$$E^+(N)[f(e^{i\theta}) - \sum_{k \geqq N} f^{\sim}(k)\varphi(k, e^{i\theta})] = 0.$$

Proof. Since $\mathscr{A} \subset \mathscr{A}_0$ we know from Theorem 5c that

$$f(e^{i\theta}) = f_N(e^{i\theta}) + \sum_{k \geqq N} f^{\sim}(k)\varphi(k, e^{i\theta})$$

where $f_N \in \mathscr{A}_0[0, N-1]$ and where the series on the right converges unconditionally in $\mathscr{A}_0$. Applying $S^+(n)$ to both sides where $n \geqq N$, and using Theorem 6a we see that if $n \geqq N$

$$S^+(n)f \cdot (e^{i\theta}) = f_N(e^{i\theta}) + \sum_{k=N}^{n} f^{\sim}(k)\varphi(k, e^{i\theta}).$$

We have

$$\lim_{n\to\infty} S^+(n)f = \lim_{n\to\infty} d(n)^{-2}u(n)E^+(0)cv(n)f - \lim_{n\to\infty} d(n)^{-2}v(n)E^+(n+1)\, cu(n)f.$$

By Corollary 3d

$$\lim_{n\to\infty} d(n)^{-2}u(n)E^+(0)cv(n)f = UE^+(0)U^{-1}f = f,$$

and

$$\lim_{n\to\infty} d(n)^{-2}v(n)E^+(n+1)cu(n)f = 0,$$

where these limits are taken in $\mathscr{A}$.

Again there is a parallel result in which $\mathscr{A}[0, \infty]$ is replaced by $\mathscr{A}[-\infty, 0]$ and the $\varphi(k)$'s by $\psi(k)$'s. We regard Theorem 6b as including both cases.

2.7. THE STRUCTURAL CONSTANTS

Theorem 7*a*. Let $\mathscr{A} \in \mathfrak{M}$ and $c \in I(\mathscr{A})$. If $D(k) \neq 0$ for $k \geqq N$ then

$$U(e^{i\theta}) = u(N, e^{i\theta}) + \sum_{k=N+1}^{\infty} \boldsymbol{\alpha}(k)d(k)\varphi(k, e^{i\theta}),$$

$$V(e^{i\theta}) = v(N, e^{i\theta}) + \sum_{k=N+1}^{\infty} \boldsymbol{\beta}(k)d(k)\psi(k, e^{i\theta}),$$

these series converging conditionally in $\mathscr{A}$.

Proof. This follows immediately from formulas (1.3.6) and (1.3.6′) and Corollary 3d. It can also be deduced from Theorem 6b.

Corollary 7b. Under the above assumptions we have

$$d(k)\boldsymbol{\alpha}(k) = \frac{1}{2\pi}\int_T U(e^{i\theta})\psi(k, e^{i\theta})c(e^{i\theta})d\theta,$$

$$d(k)\boldsymbol{\beta}(k) = \frac{1}{2\pi}\int_T V(e^{i\theta})\varphi(k, e^{i\theta})c(e^{i\theta})d\theta. \tag{2.7.1}$$

If in addition $\mathscr{A} \in \mathfrak{S}$ then $\boldsymbol{\alpha}(k) \in \boldsymbol{A}[N+1, \infty]$ and $\boldsymbol{\beta}(-k) \in \boldsymbol{A}[-\infty, -N-1]$.

Proof. The formulas (2.7.1) are evident. Suppose $\mathscr{A} \in \mathfrak{S}$; then $d(k)\boldsymbol{\alpha}(k) \in \boldsymbol{A}[N+1, \infty]$ by Lemma 5a, and since there is a constant D such that $D \geqq |d(k)^{-1}|$ for $k \geqq N+1$ it follows that $\boldsymbol{\alpha}(k) \in \boldsymbol{A}[N+1, \infty]$, etc. For another class of algebras for which a result analogous to Corollary 7b holds, see [29].

2.8. AN $\mathscr{L}^2$ FINITE SECTION INEQUALITY

Let $\mathscr{L}^2$ be the Hilbert space of all square integrable functions on T with the norm

$$||f||_2^2 = \frac{1}{2\pi}\int_T |f(e^{i\theta})|^2 \, d\theta.$$

$\mathscr{L}^2[a, b]$ is the subspace of $\mathscr{L}^2$ consisting of those functions f for which f has support in $[a, b]$. Similarly let $\mathscr{C}$ be the Banach algebra of continuous functions on T with the norm

$$||c||_\infty = \underset{\theta \in T}{\text{l.u.b.}} \, |c(e^{i\theta})|.$$

$\mathscr{C}[a, b]$ is the subspace of $\mathscr{C}$ consisting of those functions c for which c has support in $[a, b]$. $\mathscr{C}[a, b]$ is not in general a subalgebra of $\mathscr{C}$. The projections $E^+(n)$ and $E^-(n)$ are defined on $\mathscr{L}^2$ in the usual way. Note that $||E^+(n)|| = ||E^-(n)|| = 1$ where $||.||$ is the operator norm on $\mathscr{L}^2$. Let $c \in \mathscr{C}$ and define $W^+(n)$ and $W^-(n)$ as before. Then it is evident that $||W^+(n)||, ||W^-(n)|| \leqq ||c||_\infty$ for all $n = 0, 1, \ldots$.

The following result is due to Reich [58] who actually proved somewhat more.

Theorem 8a. Let

$$c(e^{i\theta}) = 1/U(e^{i\theta})V(e^{i\theta})$$

where $U, U^{-1} \in \mathscr{C}[0, \infty]$ and $V, V^{-1} \in \mathscr{C}[-\infty, 0]$. Then there exists an integer n_0 and a constant A such that if $n \geqq n_0$ we have

$$||W^+(n)^{-1}|| \leqq A, \qquad ||W^-(n)^{-1}|| \leqq A.$$

Proof. Let

$$U_m(e^{i\theta}) = \sum_{k=0}^{m-1} \left[1 - \frac{|k|}{m}\right] U(k)e^{ik\theta},$$

$$V_m(e^{i\theta}) = \sum_{k=-m+1}^{0} \left[1 - \frac{|k|}{m}\right] V(k)e^{ik\theta}.$$

By Fejer's theorem $||U_m - U||_\infty$ and $||V_m - V||_\infty$ go to zero as $m \to \infty$. Since $\mathscr{C}[0, \infty]$ and $\mathscr{C}[-\infty, 0]$ are Banach algebras Theorem 3a implies that there is an m_0 such that if $m \geqq m_0$ $U_m^{-1} \in \mathscr{C}[0, \infty]$ and $V_m^{-1} \in \mathscr{C}[-\infty, 0]$ and that $||U_m^{-1} - U^{-1}||_\infty$ and $||V_m^{-1} - V||_\infty$ go to zero as $m \to \infty$. For $m \geqq m_0$ we define $c_m = 1/U_m V_m$ and

$$W_m^+(n)f = E^+(0)E^-(n)c_m f \qquad f \in \mathscr{L}^2[0, n].$$

By Theorem 5b of I $W_m^+(n)$ is inverted by

$$Y_m^+(n)f = V_m E^-(n) U_m V_m^{-1} E^+(0) V_m f \qquad f \in \mathscr{L}^2[0, n]$$

if $n \geqq m \geqq m_0$. For $n \geqq m \geqq m_0$ we have

$$\|Y_m^+(n)\| \leqq \|V_m\|_\infty^2 \|V_m^{-1}\|_\infty \|U_m\|_\infty$$

where $\overline{\lim}_{m\to\infty} \|V_m\|_\infty^2 \|V_m^{-1}\|_\infty \|U_m\|_\infty < \infty$. Furthermore

$$\|W_m^+(n) - W^+(n)\| \leqq \|c_m - c\|_\infty$$

where $\lim_{m\to\infty} \|c_m - c\|_\infty = 0$. The remainder of the proof follows the pattern of the proof of Theorem 3b almost exactly.

The demonstrations of the following corollaries are left to the reader.

Corollary 8b. Let c and n_0 be as in Theorem 8a and let $g \in \mathscr{L}^2[0, \infty]$. If

$$f = W^+(\infty)^{-1} g, \qquad f_n = W^+(n)^{-1} E^-(n) g \qquad n \geqq n_0,$$

then there is a constant A independent of g such that

$$\|f - f_n\|_2 \leqq A \|E^+(n+1) f\|_2 \qquad n \geqq n_0.$$

Corollary 8c. Let c be as in Theorem 8a; then

$$\lim_{n\to\infty} D(n)/D(n-1) = \exp\left\{\frac{1}{2\pi}\int_T \log c(e^{i\theta}) d\theta\right\}.$$

This result, due to Reich, is a more general complex Szegö limit theorem than Corollary 3e. For a deeper study of these questions using the theory of Dirichlet algebras see Devinatz's work, in particular [20] and [22]. Devinatz's results imply as a special case that Theorem 8a and Corollaries 8b and 8c are valid under the weaker assumptions.

$$c \in \mathscr{C},$$

$$c(e^{i\theta}) \neq 0, \qquad \theta \in T,$$

$$[\arg c(e^{i\theta})]_{-\pi}^{\pi} = 0.$$

2.9. Further developments

The theory contained in sections 1–8 of this chapter holds in a much more general context. The group T may be replaced by an arbitrary locally compact Abelian group H whose dual Ξ supports a distinguished linear order '$<$' which is compatible with the group operations and for which the intersection $\{0 < \xi\} \cap B$ of the set of positive elements of Ξ with any Borel set B is again a Borel set. A detailed development of this topic can be found in [33] and [34]. It is also possible to carry out much of the preceding theory for matrix valued functions, see Bowers [11].

2.10. Szegö's theory

Let $c(d\theta)$ be a finite measure on the Borel sets of T, and let

$$c(k) = \frac{1}{2\pi}\int_T e^{-ik\theta}c(d\theta) \qquad k = \ldots, -1, 0, 1, \ldots.$$

We further assume that the support of the measure $c(d\theta)$ is infinite in order to avoid trivial cases. If $\{z(j)\}_0^n$ are arbitrary complex numbers then

$$\sum_{j,k=0}^{n} c(k-j)z(k)z(j)^* = \frac{1}{2\pi}\int_T |\sum_{j=0}^{n} z(j)e^{-ij\theta}|^2 c(d\theta) > 0$$

and thus

$$D(n) > 0 \qquad n = 0, 1, \ldots.$$

It follows that $\varphi(k, e^{i\theta})$ and $\psi(k, e^{i\theta})$ are defined for $k = 0, 1, \ldots$. Moreover since

$$\psi(k, e^{i\theta}) = \varphi(k, e^{i\theta})^*$$

the biorthogonality relations take the form

$$\frac{1}{2\pi}\int_T \varphi(k, e^{i\theta})\varphi(j, e^{i\theta})^* c(d\theta) = \delta(k, j).$$

In this case a very complete and elegant theory due initially to Szegö can be established by arguments altogether different from those employed here. Accounts of this theory can be found in the books of Szegö [65], and Grenander and Szegö [24], and a very detailed account in the book of Geronimus [23].

Recently a parallel theory for the line has been announced by Krein. For summaries of the results obtained see the notes [47], [48], [49] of Krein, and those of Achiezer [1], [2], and of Kalafati [41].

3. The Refined Szegö Limit Theorem

3.1. Introduction

It follows from Chapter 2 that if $c(e^{i\theta})$ is continuous and if

$$c(e^{i\theta}) \neq 0 \qquad \text{for} \quad \theta \in T, \quad [\arg c(e^{i\theta})]_{-\pi}^{\pi} = 0,$$

then

$$\lim_{n\to\infty} D(n)/D(n-1) = G, \tag{3.1.1}$$

where

$$D(n) = \det\,[c(k-j)]_{k,\,j=0,\ldots,n},$$

$$G = \exp\left\{\frac{1}{2\pi}\int_T \log c(e^{i\theta})d\theta\right\}.$$

In the present chapter we will show that under somewhat stronger assumptions

$$\lim_{n\to\infty} D(n)FG^{-n-1} = 1, \tag{3.1.2}$$

where

$$F = \exp\{-\sum_0^\infty c_\lambda(k)c_\lambda(-k)k\}.$$

Here

$$c_\lambda(k) = \frac{1}{2\pi}\int_T \log c(e^{i\theta})e^{-ik\theta}\,d\theta.$$

The formula (3.1.2) is a refinement of the formula (3.1.1) and for that reason, as well as because the first proof of (3.1.2) was given by Szegö, it is called the refined or strong Szegö limit theorem.

3.2. A Banach algebra method

This section is devoted to Baxter's demonstration of the refined Szegö limit theorem as later modified by the author, see [8] and [30]. Let $\boldsymbol{A}$ be the Banach algebra of all complex functions $\boldsymbol{f}$ on I for which $||\boldsymbol{f}||$ is finite where

$$||\boldsymbol{f}|| = 3\{\sum_{-\infty}^{\infty} |\boldsymbol{f}(k)| + [\sum_{-\infty}^{\infty} |\boldsymbol{f}(k)^2 k|]^{\frac{1}{2}}\}. \tag{3.2.1}$$

$\mathscr{A}$ is the corresponding algebra of functions on T. Clearly $\mathscr{A}$ is of type $\mathfrak{S}$.

Theorem 2a. If $c \in I(\mathscr{A})$, where $\mathscr{A}$ is the algebra defined by (3.2.1), then

$$\lim_{n\to\infty} FD(n)G^{-n-1} = 1.$$

Proof. From 2.3 we know that

$$\lim_{n\to\infty} \|U(e^{i\theta}) - u(n, e^{i\theta})\| = \lim_{n\to\infty} \|V(e^{i\theta}) - v(n, e^{i\theta})\| = 0. \quad (3.2.2)$$

Since U is invertible in the Banach algebra $\mathscr{A}[0, \infty]$ and V is invertible in the Banach algebra $\mathscr{A}[-\infty, 0]$ there is an integer n_0 such that if $n \geqq n_0$ $u(n)^{-1} \in \mathscr{A}[0, \infty]$ and $v(n)^{-1} \in \mathscr{A}[-\infty, 0]$. This implies that if $n \geqq n_0$ $u(n, z)$ has no zeros in $\{z \,|\, |z| \leqq 1\}$ and $v(n, z)$ has no zeros in $\{z \,|\, |z| \geqq 1\}$. Let

$$c(n, e^{i\theta}) = d(n)^2/u(n, e^{i\theta})v(n, e^{i\theta}).$$

By Szegö's identity, 1.4, we have

$$D(n)F(n)G(n)^{-n-1} = 1$$

for all sufficiently large n, where

$$G(n) = \frac{1}{2\pi} \exp\left\{\int_T \log c(n, e^{i\theta}) d\theta\right\}.$$

and where

$$F(n) = \exp\left\{\frac{1}{2\pi i}\int_{|z|=1} \log u(n, z) \frac{v(n, z)'}{v(n, z)} dz\right\}.$$

The formulas

$$\log u(n, z) = \sum_0^\infty \boldsymbol{u}_\lambda(n, k) z^k,$$

$$\log v(n, z) = \sum_{-\infty}^0 \boldsymbol{v}_\lambda(n, k) z^k,$$

are valid in some neighborhood of the unit circle. A simple computation now shows that

$$\log F(n) = -\sum_{k=0}^\infty \boldsymbol{u}_\lambda(n, k)\boldsymbol{v}_\lambda(n, -k)k,$$

$$= -\sum_{k=0}^\infty \boldsymbol{c}_\lambda(n, k)\boldsymbol{c}_\lambda(n, -k)k,$$

wher

$$\log c(n, e^{i\theta}) = \sum_{-\infty}^\infty \boldsymbol{c}_\lambda(n, k) e^{ik\theta}.$$

Since

$$\lim_{n\to\infty} \|c(e^{i\theta})-c(n, e^{i\theta})\| = 0 \tag{3.2.3}$$

it follows that

$$\lim_{n\to\infty} G(n) = G.$$

This implies, see 1.3, that for all large n

$$G/G(n) = \prod_{k=n+1}^{\infty} [1-\alpha(k)\beta(k)].$$

We know that $\alpha(k) \in A[n_0, \infty]$, $\beta(-k) \in A[-\infty, -n_0]$ see 2.7, so that

$$\sum_{n_0}^{\infty} |\alpha(k)|^2 k < \infty, \qquad \sum_{n_0}^{\infty} |\beta(k)|^2 k < \infty,$$

and therefore

$$\lim_{n\to\infty} [G/G(n)]^{n+1} = 1.$$

Since $c^{-1} = UVd^{-2}$, c is invertible in $\mathscr{A}$, and the limit relation (3.2.3) is equivalent to

$$\lim_{n\to\infty} \|1-c(n, e^{i\theta})c(e^{i\theta})^{-1}\| = 0.$$

Using $\log x = -\sum_1^{\infty} (1-x)^k/k$ it is now easily seen that

$$\lim_{n\to\infty} \|\log c(n, e^{i\theta})-\log c(e^{i\theta})\| = 0,$$

and a fortiori that

$$\lim_{n\to\infty} \sum_{k=-\infty}^{\infty} |c_\lambda(n, k)-c_\lambda(k)|^2 |k| = 0.$$

Consequently

$$\lim_{n\to\infty} F(n) = F$$

and our proof is complete.

If in addition to the assumptions of Theorem 1a we suppose that $D(k) \neq 0$ for $k = 0, 1, 2, \ldots$ then it follows from Corollary 3c and Lemma 4a of Chapter 1 that

$$F(c) = \prod_{k=1}^{\infty} [1-\alpha(k)\beta(k)]^k,$$

$$G(c) = c(0) \prod_{k=1}^{\infty} [1-\alpha(k)\beta(k)].$$

We recall from 2.2 that $c \in I(\mathscr{A})$ if and only if

$$\begin{aligned} &c \in \mathscr{A} \\ &c(e^{i\theta}) \neq 0 \qquad \theta \in T, \\ &[\arg c(e^{i\theta})]_{-\pi}^{\pi} = 0. \end{aligned} \tag{3.2.4}$$

Consequently Theorem 2a holds if the assumption $c \in I(\mathscr{A})$ is replaced by the assumptions (3.2.4).

It is interesting to consider a somewhat different Banach algebra. Let $\boldsymbol{A}$ be the Banach algebra of all complex functions for which $\lim_{k\to -\infty} kf(k) = 0$ and $\|f\| < \infty$ where

$$\|f\| = 2\{\sum_{k=-\infty}^{\infty} |f(k)| + \operatorname{l.u.b.}_{k\leq 0} |f(k)k|\}.$$

Exactly the same arguments used above lead to the following result.

Theorem 2b. If $c \in I(\mathscr{A})$ where $\mathscr{A}$ is the algebra defined above then

$$\lim_{n\to\infty} FD(n)G^{-n-1} = 1.$$

Note that the assumptions of Theorems 2a and 2b are incomparable, neither implying the other. Theorem 2b generalizes a result obtained by Baxter and Schmidt in [9].

3.3. Devinatz's method

Quite recently an improved version of Theorem 2a has been obtained by Devinatz [21] using rather different methods.

Lemma 3a. If $f(e^{i\theta}) \in \mathscr{L}^2(T)$ then

$$\frac{1}{4\pi^2}\int_T\int_T \left|\frac{f(e^{i\theta})-f(e^{i\varphi})}{e^{i\theta}-e^{i\varphi}}\right|^2 d\theta\, d\varphi = \sum_{k=-\infty}^{\infty} |f(k)|^2|k|.$$

Proof. By Parseval's equality and Fubini's theorem we have

$$\sum_{-\infty}^{\infty} |(1-e^{ik\varphi})f(k)|^2 = \frac{1}{2\pi}\int_T |f(e^{i\theta})-f(e^{i\theta+i\varphi})|^2\, d\theta,$$

$$\sum_{-\infty}^{\infty} |f(k)|^2 \frac{1}{2\pi}\int_T \left|\frac{1-e^{ik\varphi}}{1-e^{i\varphi}}\right|^2 d\varphi = \frac{1}{4\pi^2}\int_T\int_T \left|\frac{f(e^{i\theta})-f(e^{i\theta+i\varphi})}{1-e^{i\varphi}}\right|^2 d\theta\, d\varphi.$$

Changing variables on the right and using the evaluation

$$\frac{1}{2\pi}\int_T \left|\frac{1-e^{ik\varphi}}{1-e^{i\varphi}}\right|^2 d\varphi = \frac{1}{2\pi}\int_T |\sum_{j=0}^{k-1} e^{ij\varphi}|^2 d\varphi = k$$

on the left we are done.

For any function $f(e^{i\theta})$ in $\mathscr{L}^2(T)$ let us set

$$\mathscr{N}[f] = \{\sum_{-\infty}^{\infty} |f(k)|^2|k|\}^{\frac{1}{2}}.$$

We define $\mathscr{N}$ to be the set of all f in $\mathscr{L}^2(T)$ for which $\mathscr{N}[f]$ is finite.

Corollary 3b. Let $\Omega(z)$ be a complex valued function defined on a compact set F in the z-plane and such that $|\Omega(z_1)-\Omega(z_2)| \leqq B|z_1-z_2|$. If $f \in \mathscr{N}$ and if the range of f lies in F then $\mathscr{N}[g] \leqq B\mathscr{N}[f]$, where $g(e^{i\theta}) = \Omega[f(e^{i\theta})]$. A fortiori $g \in \mathscr{N}$.

This result plays the same role here that the Wiener-Lévy theorem played in 3.2.

We fix a function $c(e^{i\theta})$, $\theta \in T$, with the following properties:

i. c is continuous on T;
ii. $c \neq 0$ for $\theta \in T$;
iii. $[\arg c]_{-\pi}^{\pi} = 0$; (3.3.1)
iv. $c \in \mathscr{N}$;
v. $E^+(0) \log c$ and $E^- \log c$ are continuous.

Lemma 3c. Let c satisfy the conditions (3.3.1) Let $U_\lambda = -E^+(0) \log c$, $V_\lambda = -E^-(0) \log c$, $U = \exp U_\lambda$, $V = \exp V_\lambda$, and $d^2 = U(0) = V(0)$. Then U_λ, V_λ, U, and V belong to $\mathscr{N}$ and are continuous, and $c = d^2/UV$.

Proof. Choose a trigonometric polynomial $p(e^{i\theta})$ which does not vanish and is such that

$$|\log c(e^{i\theta})+p(e^{i\theta})| \leqq \frac{\pi}{2} \qquad \theta \in T.$$

Let

$$m_1 = \inf_T |ce^p|, \qquad m_2 = \sup_T |ce^p|,$$

and set

$$F = \{z: m_1 \leqq |z| \leqq m_2, Rlz \geqq 0\}, \qquad \Omega(z) = \log z.$$

It follows from Corollary 3b that $\log ce^p \in \mathcal{N}$. Since $p \in \mathcal{N}$ trivially and since

$$\mathcal{N}[\log c] \leqq \mathcal{N}[p+\log c]+\mathcal{N}[p]$$

we see that $\log c \in \mathcal{N}$. A fortiori U_λ, $V_\lambda \in \mathcal{N}$, etc.

Theorem 3d. If $c(e^{i\theta})$ satisfies assumptions i.–v. of (3.3.1) then

$$\lim_{n\to\infty} D(n)G^{-n-1}F = 1.$$

Proof. It follows from Lemma 3c and Theorem 8a of Chapter 2 that there exists a non-negative integer n_0 and a constant M such that

$$\|W^+(n)f\|_2 \leqq M\|f\|_2, \qquad \|W^-(n)g\|_2 \leqq M\|g\|_2,$$

for $n \geqq n_0$, and $f \in \mathcal{L}^2[0, n]$, $g \in \mathcal{L}^2[-n, 0]$. Thus

$$u(n, e^{i\theta}) = W^+(n)^{-1}1 \quad \text{and} \quad v(n, e^{i\theta}) = W^-(n)^{-1}1$$

are well defined for $n \geqq n_0$. As in 3.2, it is sufficient to prove that $[G(n)/G]^{n+1} \to 1$ and $F(n) \to F$ as $n \to \infty$. Since $u(n)c = g_1+1+g_2$ where $g_1 \in \mathcal{L}^2[-\infty, -1]$, $g_2 \in \mathcal{L}^2[n+1, \infty]$ and since $v(n) \in \mathcal{L}^2[-n, 0]$ we see that

$$\frac{1}{2\pi}\int_T \frac{u(n, e^{i\theta})}{\boldsymbol{u}(n, 0)}\, c(e^{i\theta})\, \frac{v(n, e^{i\theta})}{\boldsymbol{v}(n, 0)}\, d\theta = \frac{1}{\boldsymbol{v}(n, 0)} = G(n).$$

Similarly

$$\frac{1}{2\pi}\int_T \frac{u(n, e^{i\theta})}{\boldsymbol{u}(n, 0)}\, \frac{d^2}{V(0)u(e^{i\theta})}\, d\theta = G.$$

Using $c = d^2/UV$ we find that

$$\begin{aligned} G(n)-G &= \frac{d^2}{2\pi}\int_T \frac{u(n, e^{i\theta})}{\boldsymbol{u}(n, 0)U(e^{i\theta})}\left\{\frac{v(n, e^{i\theta})}{\boldsymbol{v}(n, 0)V(e^{i\theta})} - \frac{1}{V(0)}\right\} d\theta, \\ &= \frac{d^2}{2\pi}\int_T \left\{\frac{u(n, e^{i\theta})}{\boldsymbol{u}(n, 0)U(e^{i\theta})} - \frac{1}{U(0)}\right\}\left\{\frac{v(n, e^{i\theta})}{\boldsymbol{v}(n, 0)V(e^{i\theta})} - \frac{1}{V(0)}\right\} d\theta. \end{aligned}$$

By Schwarz's inequality

$$|G-G(n)| \leqq d^2 \left\|\frac{u(n, e^{i\theta})}{\boldsymbol{u}(n, 0)U(e^{i\theta})} - \frac{1}{U(0)}\right\|_2 \left\|\frac{v(n, e^{i\theta})}{\boldsymbol{v}(n, 0)V(e^{i\theta})} - \frac{1}{V(0)}\right\|_2.$$

Using Corollary 8b of Chapter 2 we see that

$$\|u(n)-U\|_2 \leqq M\Big[\sum_{n+1}^{\infty} |U(k)|^2\Big]^{\frac{1}{2}} \leqq M\left[\frac{1}{n}\sum_{n+1}^{\infty} |U(k)|^2 k\right]^{\frac{1}{2}} = o(n^{-\frac{1}{2}})$$

and therefore, a fortiori,

$$|\boldsymbol{u}(n, 0) - \boldsymbol{U}(0)| = o(n^{-\frac{1}{2}}).$$

Similar results hold for the 'v's', and from this it follows that

$$|G - G(n)| = o(n^{-1}) \quad \text{as} \quad n \to \infty.$$

Since

$$\frac{G(n)}{G} = 1 - \frac{G - G(n)}{G} = 1 + o(n^{-1})$$

it follows that $[G(n)/G]^{n+1} \to 1$ as $n \to \infty$, as we wished to show.

It remains to prove that $F(n) \to F$ as $n \to \infty$. We begin by proving that $||U - u(n)||_\infty$ and $||V - v(n)||_\infty$ go to zero as n goes to ∞. Let

$$U(n, e^{i\theta}) = \sum_{0}^{n} \left(1 - \frac{|k|}{n+1}\right) U(k)e^{ik\theta},$$

$$V(n, e^{i\theta}) = \sum_{-n}^{0} \left(1 - \frac{|k|}{n+1}\right) V(k)e^{ik\theta}.$$

It is easy to see that

$$W^+(n)[U(n) - u(n)] = E^+(0)E^-(n)\{c[U(n) - U]\},$$

which implies that

$$||W^+(n)[U(n) - u(n)]||_2^2 \leqq \frac{1}{2\pi} \int_T |c[U(n) - U]|^2 \, d\theta.$$

If $m = \sup |c(e^{i\theta})|$ on T then

$$\begin{aligned} ||W^+(n)[U(n) - u(n)]||_2^2 &\leqq \frac{m}{2\pi} \int_T |U(n) - U|^2 \, d\theta, \\ &\leqq \sum_{k=0}^{n} \left(\frac{k}{n+1}\right)^2 |\boldsymbol{U}(k)|^2 + \sum_{k+1}^{\infty} |\boldsymbol{U}(k)|^2, \\ &= o(n^{-1}) \qquad \text{as} \quad n \to \infty. \end{aligned}$$

It follows, see 2.8, that

$$||U(n) - u(n)||_2 = o(n^{-\frac{1}{2}}) \qquad \text{as} \quad n \to \infty.$$

By Schwarz's inequality

$$\begin{aligned} ||U(n) - u(n)||_\infty &\leqq \sum_{k=0}^{n} |\boldsymbol{U}(n, k) - \boldsymbol{u}(n, k)|, \\ &\leqq n^{\frac{1}{2}} [\sum_{k=0}^{n} |\boldsymbol{U}(n, k) - \boldsymbol{u}(n, k)|^2]^{\frac{1}{2}} \\ &= o(1) \qquad \text{as} \quad n \to \infty. \end{aligned} \tag{3.3.2}$$

By Fejërs' theorem, since U is continuous,

$$\|U - U(n)\|_\infty = o(1) \qquad \text{as} \quad n \to \infty. \tag{3.3.3}$$

Combining (3.3.2) and (3.3.3) we have

$$\|U - u(n)\|_\infty \to 0 \qquad \text{as} \quad n \to \infty. \tag{3.3.4}$$

By Corollary 8b of Chapter 2 it follows that

$$\sum_{k=0}^{n} |U(k) - u(n, k)|^2 \leqq M \sum_{n+1}^{\infty} |U(k)|^2$$

which implies that

$$\begin{aligned}\sum_{k=0}^{\infty} |U(k) - u(n, k)|^2 |k| &\leqq M_1 \sum_{n+1}^{\infty} |U(k)|^2 |k| \\ &= o(1) \qquad \text{as} \quad n \to \infty.\end{aligned}$$

It follows from (3.3.4) that $u(n, e^{i\theta})/U(e^{i\theta}) \to 1$ uniformly for $\theta \in T$ as $n \to \infty$. We can therefore apply Corollary 2b with $\Omega(z) = \log z$ to show that for some constant A and all large n

$$\begin{aligned}\sum_{k=0}^{\infty} |u_\lambda(n, k) - U_\lambda(k)|^2 |k| &\leqq A \sum_{k=0}^{\infty} |u(n, k) - U(k)|^2 |k| \\ &= o(1) \qquad \text{as} \quad n \to \infty.\end{aligned}$$

Similarly

$$\sum_{k=-\infty}^{0} |v_\lambda(n, k) - V_\lambda(k)|^2 |k| = o(1) \qquad \text{as} \quad n \to \infty,$$

and we are done.

3.4. Kac's method

There is an extremely interesting alternate approach to the refined Szegö limit theorem due to Kac. Let $\{\lambda(n, k)\}_{k=1}^{n+1}$ be the eigen values of

$$[c(k-j)]_{k, j=0, \ldots, n}$$

where we assume that

$$\sum_{-\infty}^{\infty} |c(k)| = B < \infty$$

$$\sum_{-\infty}^{\infty} |c(k)|^2 |k| = C^2 < \infty.$$

A simple calculation, see 4.1, shows that

$$\sum_{k=1}^{n+1} \lambda(n, k)^s - \frac{n+1}{2\pi} \int_T c(e^{i\theta})^s d\theta \tag{3.4.1}$$

$$= \sum_{j^{(s)}} R(j^{(s)})_{n+1} c(j_1) \ldots c(j_s),$$

where $j^{(s)} = (j_1, j_2, \ldots, j_s)$ and only s-tuples $j^{(s)}$ are considered for which $j_1 + j_2 + \ldots + j_s = 0$, and where

$$R(j^{(s)}) = \max(0, j_1, j_1+j_2, \ldots, j_1 + \ldots + j_{s-1}) - \min(0, j_1, j_1+j_2, \ldots, j_1 + \ldots + j_{s-1}),$$

$$R(j^{(s)})_{n+1} = \min\{R(j^{(s)}), n+1\}.$$

It can be shown that

$$\sum_{j^{(s)}} |R(j^{(s)}) c(j_1) \ldots c(j_s)| \leqq \begin{cases} 0 & s = 1 \\ s^2 C^2 B^{s-2} & s \geqq 2. \end{cases}$$

Choosing any complex ξ for which $B|\xi| < 1$, multiplying (3.4.1) by $\xi^s s^{-1}$, summing over s from 1 to ∞, and letting $n \to \infty$ we find that, in an evident notation,

$$\lim_{n \to \infty} \{\log D[n, 1 - \xi c] - (n+1) \log G[1 - \xi c]\}$$

$$= -\sum_{s=1}^{\infty} \frac{\xi^s}{s} \sum_{j^{(s)}} R(j^{(s)}) c(j_1) \ldots c(j_s). \tag{3.4.2}$$

Using a combinatorial identity of Hunt it is possible to show that

$$\sum_{s=1}^{\infty} \frac{\xi^s}{s} \sum_{j^{(s)}} R(j^{(s)}) c(j_1) \ldots c(j_s) = \log F[1 - \xi c].$$

See [37], and also [38], for the details of these arguments. Renaming $1 - \xi c$ as c we obtain the following.

Theorem 4a. If

$$\sum_{-\infty}^{\infty} |c(k)| < \infty, \qquad \sum_{-\infty}^{\infty} |c(k)|^2 |k| < \infty,$$

and if

$$|c(0)| > \sum_{k \neq 0} |c(k)|, \tag{3.4.3}$$

then

$$\lim_{n\to\infty} D(n)G^{-n-1}F = 1.$$

The assumption (3.4.3) implies

$$c(e^{i\theta}) \neq 0 \qquad \theta \in T, \qquad [\arg c(e^{i\theta})]_{-\pi}^{\pi} = 0. \tag{3.4.4}$$

Indeed (3.4.3) is much stronger than (3.4.4). Recently the author has shown how starting from Theorem 4a it is possible to obtain the results of section 2. This derivation is of particular interest in that no use is made of Szegö's identity and consequently these arguments can be applied to some cases (notably Toeplitz matrices generated from ultraspherical polynomials) in which no analogue of Szegö's identity exists. See Davis and Hirschman [16] and Hirschman [32].

See also in these connections the papers of Widom [72] and Mejlbo and Schmidt [52].

3.5. A limit theorem for the line

We will describe here the analogue for the line of the refined Szegö limit theorem for the circle. Since we will work with measures on the line which are absolutely continuous except at $\xi = 0$, it will be convenient to represent them as generalized functions

$$\boldsymbol{f}(\xi) = \boldsymbol{f}_s\delta(\xi)+\boldsymbol{f}_a(\xi),$$

where $\boldsymbol{f}_s$ is a complex constant, $\delta(\xi)$ is the Dirac delta function, and $\boldsymbol{f}_a(\xi)$ is a complex function integrable on a suitable interval. Let

$$\boldsymbol{c}(\xi) = \delta(\xi)-\boldsymbol{c}_a(\xi) \qquad -\infty < \xi < \infty$$

be a given generalized function where $\boldsymbol{c}_a(\xi)$ is locally integrable. Note that in this one case we write $-\boldsymbol{c}_a$ instead of $\boldsymbol{c}_a$. The analogue for the line of $W^+(n)\boldsymbol{f} = \boldsymbol{g}$ for the circle is

$$\boldsymbol{W}^+(\eta)\boldsymbol{f} = \boldsymbol{E}^+(0)\boldsymbol{E}^-(\eta)\boldsymbol{c} * \boldsymbol{f} = \boldsymbol{g} \tag{3.5.1}$$

where $\boldsymbol{f}_a(\xi) \in L^1[0, \eta]$. This is equivalent to

$$\boldsymbol{f}_s = \boldsymbol{g}_s,$$

together with

$$\boldsymbol{f}_a(\xi)-\int_0^\eta \boldsymbol{c}_a(\xi-\eta)\boldsymbol{f}_a(\eta)\mathrm{d}\eta = \boldsymbol{g}_a(\xi)-\boldsymbol{g}_s\boldsymbol{c}_a(\xi).$$

See in these connections 2.10 and the papers referred to there. Let us now assume that

$$\int_{-r}^{r} |c_a(\xi)|^2 \mathrm{d}\xi < \infty$$

for every $r > 0$. Appealing to the theory of Fredholm integral equations [61], we see that (3.5.1) is invertible if and only if $D(\eta) \neq 0$ where $D(\eta)$ is the Fredholm determinant.

$$D(\eta) = 1 + \sum_{n=1}^{\infty} \frac{(-1)^n}{n!} \int_0^\eta \cdots \int \begin{vmatrix} c_a(\xi_1-\xi_1) & \dots & c_a(\xi_1-\xi_n) \\ \vdots & & \vdots \\ c_a(\xi_n-\xi_1) & \dots & c_a(\xi_n-\xi_n) \end{vmatrix} \mathrm{d}\xi_1 \dots \mathrm{d}\xi_n .$$

We now make the assumptions

i. $\int_{-\infty}^{\infty} |c_a(\xi)| \mathrm{d}\xi < \infty, \quad \int_{-\infty}^{\infty} |c_a(\xi)^2| (1+|\xi|) \mathrm{d}\xi < \infty,$

ii. $c(t) \neq 0 \quad -\infty < t < \infty, \quad [\arg c(t)]_{-\infty}^{\infty} = 0,$

where

$$c(t) = \int_{-\infty}^{\infty} c(\xi) e^{i\xi t} = 1 - \int_{-\infty}^{\infty} c_a(\xi) e^{i\xi t} \mathrm{d}\xi .$$

We further assume the existence of the limits

iii. $\lim_{h\to 0+} h^{-1} \int_0^h c_a(\xi) \mathrm{d}\xi \quad \text{and} \quad \lim_{h\to 0+} h^{-1} \int_{-h}^{0} c_a(\xi) \mathrm{d}\xi,$

and we assume throughout that

$$c_a(0) = \lim_{h\to 0+} (2h)^{-1} \int_{-h}^{h} c_a(\xi) \mathrm{d}\xi .$$

This normalization does not alter the Toeplitz operator $W^+(\eta)$ and consequently involves no loss of generality.

We define

$$G = \exp\left\{ \lim_{T\to\infty} \frac{1}{2\pi} \int_{-T}^{T} \left(1 - \frac{t^2}{T^2}\right)^2 \log c(t) \mathrm{d}t \right\},$$

and

$$F = \exp\left\{ -\int_0^\infty c_a^\lambda(\xi) c_a^\lambda(-\xi) \xi \mathrm{d}\xi \right\},$$

where

$$\log c(t) = \int_{-\infty}^{\infty} c_a^{\lambda}(\xi) e^{it\xi} d\xi.$$

Theorem 5a. If the assumptions i.-iii. are satisfied, then

$$\lim_{\eta \to \infty} D(\eta) F G^{-\eta} = 1.$$

This formula was first obtained by Kac in 1954, [37], who derived it using his combinatorial method, see 3.4, but under rather severe assumptions. The subject was taken up again in 1964 by Achiezer [1]. Starting from Kac's assumptions he derived Theorem 4a, although with a different formula for F. The result stated above is due to the author [35]. The proof follows very closely the pattern of 3.2. The principal difficulty lies in the derivation of the analogue for the line of Szegö's identity for the circle. Professor Devinatz has informed me that his method can also be applied here to obtain a still more general result.

3.6. Further developments

Recent work in quantum statistical mechanics, see Lenard [51], has led to the need for rather precise estimates for $D(n)$ as $n \to \infty$ under hypotheses on c weaker than those in 3.2 and 3.3. Consider, for example, the special case

$$c(e^{i\theta}) = 1 - \cos \theta.$$

Since $c(e^{i\theta}) = 0$ for $\theta = 0$ the familiar forms of the refined Szegö limit theorem do not apply. However, direct calculation in this case gives

$$D(n) = (\tfrac{1}{2})^{n+1}(n+2),$$

$$G = \tfrac{1}{2},$$

from which it follows that

$$\log D(n) - (n+1) \log G \sim \log n \quad \text{as} \quad n \to \infty. \tag{3.6.1}$$

Under certain circumstances probability arguments based on (3.4.2) can be applied to obtain formulas of this kind, see Kac [39]. Alternatively, valuable inequalities can be derived by approximations using functions for which the refined Szegö theorem is valid; see

Szegö's results in [51]. However, these problems are not completely understood at the present time.

Other important asymptotic properties of Toeplitz matrices have been derived from studies of homogeneous random walks on the integers in the presence of absorbing barriers. However, these results involve rather deep considerations from probability theory and are difficult to explain out of context. The interested reader should consult the papers of Spitzer and Stone [63], Widom [74], Watanabe [70], and Kesten [42], [43], and also Spitzer's book [62].

4. Eigen Values of Hermitian Toeplitz Operators

4.1. Introduction

Let $\boldsymbol{L}^2[0, n]$ be the finite Hilbert space of all complex functions $\boldsymbol{f}(k)$ on $k = 0, 1, \ldots, n$ with inner product $(\boldsymbol{f}|\boldsymbol{g}) = \sum_0^n \boldsymbol{f}(k)\boldsymbol{g}(k)^*$ and norm $||f|| = (\boldsymbol{f}|\boldsymbol{f})^{\frac{1}{2}}$. If $c \in Rl\mathscr{L}^1(T)$ the operator

$$\boldsymbol{W}^+[c, n]\boldsymbol{f} \cdot (k) = \sum_{j=0}^{n} c(k-j)\boldsymbol{f}(j) \qquad \boldsymbol{f} \in \boldsymbol{L}^2[0, n]$$

is self-adjoint; equivalently the matrix

$$M[c, n] = [c(k-j)]_{k,\, j=0,\, 1, \ldots,\, n}$$

is Hermitian. Let $\lambda(n, 1) \leqq \lambda(n, 2) \leqq \ldots \leqq \lambda(n, n+1)$ be the necessarily real eigen values of $\boldsymbol{W}^+[c, n]$. Great interest has centered on the problems of describing the behaviour of the $\{\lambda(n, k)\}_{k=1}^{n+1}$ as $n \to \infty$, and in recent years very considerable progress has been made in this direction. The ideas involved in these developments are altogether different from those discussed in the preceding chapters, and they are rather technical. Consequently I have limited myself here to stating a few typical theorems and to indicating how these are proved, without, however, carrying out the proofs in detail. The reader who wishes to familiarize himself with the wide variety of results of this kind that exist (as well as their demonstrations) should consult the references given.

The simplest and most fundamental result in this direction is the global distribution formula of Szegö' Although this result is classical its importance for our discussions is such that it has seemed best to include its demonstration.

For $c(e^{i\theta}) \in Rl\mathscr{L}^1(T)$ we define the measure

$$\alpha(E) = \frac{1}{2\pi}\int_{R(E)} d\theta \tag{4.1.1}$$

where $R(E) = \{\theta \in T | c(e^{i\theta}) \in E\}$. Here E is any Borel set on the real line. Similarly for each $n = 0, 1, \ldots$ set

$$\alpha(n, E) = \frac{1}{n+1}\sum_{k \in N_n(E)} 1 \tag{4.1.2}$$

where $N_n(E) = \{k \in [1, n+1] | \lambda(n, k) \in E\}$. It is evident that both $\alpha(\cdot)$ and $\alpha(n, \cdot)$ are positive measures on the Borel sets of the line with total mass 1. Szegö's result is the following.

Theorem 1a. Let $c \in RlL^1(T)$ and let the measures $\alpha(\cdot)$ and $\alpha(n, \cdot)$ be defined as above. Then

$$\alpha(n, \cdot) \rightharpoonup \alpha(\cdot) \qquad \text{as} \quad n \to \infty; \tag{4.1.3}$$

that is, if $f(\lambda)$ is any continuous function on the line with compact support then

$$\lim_{n\to\infty}\int_{-\infty}^{\infty} f(\lambda)\alpha(n, d\lambda) = \int_{-\infty}^{\infty} f(\lambda)\alpha(d\lambda).$$

Proof. Suppose first that $\sum_{-\infty}^{\infty} |c(k)| < \infty$. We then have

$$\sum_{i=1}^{n+1} \lambda(n, i)^s = \operatorname{tr}\{M[n, c]^s\} = \sum_{\substack{0 \leqq k_\nu \leqq n \\ \nu=1,\ldots,s}} c(k_1-k_2)c(k_2-k_3)\ldots c(k_s-k_1),$$

$$\int_{-\infty}^{\infty} \lambda^s\alpha(n, d\lambda) = \sum_{j_1+\ldots+j_s=0} (n+1)^{-1}S(n, j^{(s)})c(j_1)\ldots c(j_s),$$

where $j^{(s)} = (j_1, j_2, \ldots, j_s)$ and where $S(n, j^{(s)})$ is the number of solutions of the system of equations

$$k_1-k_2 = j_1,\ k_2-k_3 = j_2, \ldots, k_s-k_1 = j_s$$

under the restrictions $0 \leqq k_\nu \leqq n$, $\nu = 1, \ldots, s$. It is easy to see that

$$S(n, j^{(s)}) = n+1-\min\{R(j^{(s)}), n+1\}$$

where

$$\begin{aligned} R(j^{(s)}) = \max(0, j_1, j_1+j_2, \ldots, j_1+\ldots+j_s) \\ -\min(0, j_1, j_1+j_2, \ldots, j_1+\ldots+j_s). \end{aligned}$$

This implies that

$$n+1 \geqq S(n, j^{(s)}) \geqq n+1-(|j_1|+\ldots+|j_s|),$$

and thus that

$$\lim_{n\to\infty}\int_{-\infty}^{\infty}\lambda^s\alpha(n, \mathrm{d}\lambda) = \frac{1}{2\pi}\int_T c(\mathrm{e}^{i\theta})^s \mathrm{d}\theta = \int_{-\infty}^{\infty}\lambda^s\alpha(\mathrm{d}\lambda).$$

Since supp $\alpha(n, .) \subset [-\|c\|_0, \|c\|_0]$ it follows from the Weierstrass approximation theorem that

$$\lim_{n\to\infty}\int_{-\infty}^{\infty} f(\lambda)\alpha(n, \mathrm{d}\lambda) = \int_{-\infty}^{\infty} f(\lambda)\alpha(\mathrm{d}\lambda)$$

for every continuous function $f(\lambda)$ with compact support, and our assertion is thus proved in the present case.

Suppose next that $c \in Rl\mathscr{L}^1(T)$ is such that $c(e^{i\theta}) = \lim_{r\to\infty} c(r, e^{i\theta})$ for almost all $\theta \in T$ where:

i. $c(r, \mathrm{e}^{i\theta}) \in Rl\mathscr{L}^1(T)$ for $r = 1, 2, \ldots$

and (4.1.3) holds for each $c(r, e^{i\theta})$ $r = 1, 2, \ldots$;

ii. $c(1, \mathrm{e}^{i\theta}) \leqq c(2, \mathrm{e}^{i\theta}) \leqq c(3, \mathrm{e}^{i\theta}) \leqq \ldots$.

Let $\{\lambda(r, n, i)\}_{i=1}^{n+1}$ be the eigen values of $\boldsymbol{W}^+[c(r), n]$ in non-decreasing order. Let $\alpha(r, n, \mathrm{d}\lambda)$ be the measure corresponding to the $\lambda(r, n, i)$'s and $\alpha(r, \mathrm{d}\lambda)$ the measure corresponding to $c(r, e^{i\theta})$. For $\boldsymbol{f} \in \boldsymbol{L}^2[0, n]$ let $f(e^{i\theta}) = \sum_0^n \boldsymbol{f}(k)e^{ik\theta}$. Since $c(r, e^{i\theta}) \leqq c(e^{i\theta})$ for almost all $\theta \in T$ we have

$$\begin{aligned}(\boldsymbol{W}^+[c(r), n]\boldsymbol{f}|\boldsymbol{f}) &= \frac{1}{2\pi}\int_T c(r, \mathrm{e}^{i\theta})|f(\mathrm{e}^{i\theta})|^2 \mathrm{d}\theta \\ &\leqq \frac{1}{2\pi}\int_T c(\mathrm{e}^{i\theta})|f(\mathrm{e}^{i\theta}|^2 \mathrm{d}\theta = (\boldsymbol{W}^+[c, n]\boldsymbol{f}|\boldsymbol{f}).\end{aligned}$$

It follows from the Weyl-Courant lemma, see [13, p. 33], that

$$\lambda(r, n, i) \leqq \lambda(n, i) \qquad i = 1, \ldots, n+1.$$

Thus

$$\begin{aligned}(n+1)^{-1}\sum_{i=1}^{n+1}|\lambda(n, i)-\lambda(r, n, i)| &= (n+1)^{-1}\sum_{i=1}^{n+1}[\lambda(n, i)-\lambda(r, n, i)] \\ &= \frac{1}{2\pi}\int_T\{c(\mathrm{e}^{i\theta})-c(r, \mathrm{e}^{i\theta})\}\mathrm{d}\theta.\end{aligned}$$

We can now apply the Lebesgue limit theorem to show that

$$\lim_{r\to\infty}(n+1)^{-1}\sum_{i=1}^{n+1}|\lambda(n,i)-\lambda(r,n,i)| = 0.$$

It follows that if $f(\lambda)$ is any real continuous function on the line with compact support then given $\varepsilon > 0$ there exists r_1 such that

$$\left|\int_{-\infty}^{\infty} f(\lambda)\alpha(n,\mathrm{d}\lambda)-\int_{-\infty}^{\infty} f(\lambda)\alpha(r,n,\mathrm{d}\lambda)\right| < \varepsilon \qquad r \geqq r_1$$

and all $n = 1, 2, \ldots$. Moreover there obviously exists r_2 such that

$$\left|\int_{-\infty}^{\infty} f(\lambda)\alpha(\mathrm{d}\lambda)-\int_{-\infty}^{\infty} f(\lambda)\alpha(r,\mathrm{d}\lambda)\right| < \varepsilon \qquad r \geqq r_2.$$

Since by assumption

$$\lim_{n\to\infty}\int_{-\infty}^{\infty} f(\lambda)\alpha(r,n,\mathrm{d}\lambda) = \int_{-\infty}^{\infty} f(\lambda)\alpha(r,\mathrm{d}\lambda)$$

it follows that

$$\overline{\lim_{n\to\infty}}\left|\int_{-\infty}^{\infty} f(\lambda)\alpha(n,\mathrm{d}\lambda)-\int_{-\infty}^{\infty} f(\lambda)\alpha(\mathrm{d}\lambda)\right| < 2\varepsilon,$$

and hence that $\alpha(n,.)$ converges weakly to $\alpha(.)$ as $n \to \infty$. Thus (4.1.3) holds for $c(e^{i\theta})$. Assumption ii. may evidently be replaced by

ii′. $c(1, \mathrm{e}^{i\theta}) \geqq c(2, \mathrm{e}^{i\theta}) \geqq c(3, \mathrm{e}^{i\theta}) \geqq \ldots$.

Since the smallest subclass of $Rl\mathscr{L}^1(T)$ which is closed under monotone limits and which contains all real functions in $\mathscr{A}_0$ is $Rl\mathscr{L}^1(T)$ itself our theorem follows. This proof is due to Grenander and Szegö [24, § 7.5].

Corollary 1b. If the Lebesgue measure of the sets $\{\theta|c(e^{i\theta}) = a\}$ and $\{\theta|c(e^{i\theta}) = b\}$ is zero then

$$\lim_{n\to\infty}(n+1)^{-1}N(n; a, b) = (2\pi)^{-1}M(a, b),$$

where $N(n; a, b)$ is the number of $\lambda(n, i)$'s in $[a, b]$, and $M(a, b)$ is the measure of the set $\{\theta|a \leqq c(e^{i\theta}) \leqq b\}$.

Global distribution theorems can be shown to hold in very general contexts. See Grenander and Szegö [24] and Krieger [50].

Scholium 1c. Let $c(e^{i\theta}) \in Rl\mathscr{L}^1(T)$ and let

$$m_1 = \operatorname*{ess\,inf}_{\theta\in T} c(e^{i\theta}), \qquad m_2 = \operatorname*{ess\,sup}_{\theta\in T} c(e^{i\theta})$$

so that m_1 is finite or $-\infty$, and m_2 is finite or $+\infty$. If $m_1 < m_2$ then

$$m_1 < \lambda(n, 1) < \lambda(n, 2) < \ldots < \lambda(n, n+1) < m_2$$

for $n = 1, 2, \ldots$.

Proof. That $m_1 < \lambda(n, 1)$ follows from

$$(W^+[c, n]f|f) = \frac{1}{2\pi}\int_T c(e^{i\theta})|f(e^{i\theta})|^2\,d\theta$$

$$> \frac{1}{2\pi}\int_T m_1|f(e^{i\theta})|^2\,d\theta = m_1\|f\|^2,$$

etc.

If $c \in \mathscr{L}^1(T)$ but is not real the operators $W^+[c, n]$ are no longer self-adjoint so that the preceding methods do not apply, and in fact little is known except when c is a trigonometric polynomial. A complete solution in this case has been given by Schmidt and Spitzer [59], and Hirschman [36].

4.2. Minimum eigen values

Let $c(e^{i\theta})$ be real and continuous for $\theta \in T$ and not a constant. It follows from 4.1 that for k fixed and n large

$$\lambda(n, k) \geqq \lambda(n, k-1) \geqq \ldots \geqq \lambda(n, 1) > \min_T c(e^{i\theta}),$$

and also that

$$\lim_{n\to\infty} \lambda(n. k) = \min_T c(e^{i\theta}).$$

In this section and the next we will be concerned with determining how, for k fixed, $\lambda(n, k)$ approaches min c as n goes to ∞.

Since the eigen values of $M[c(e^{i\theta})+a, n]$ are $\{\lambda(n, i)+a\}_{i=1}^{n+1}$, we may assume without loss of generality that min $c = 0$, and since the eigen values of $M[c(e^{i\theta-\theta_0}), n]$ are $\{\lambda(n, i)\}_{i=1}^{n+1}$ we may further assume that $c(e^{i\theta}) = 0$ when $\theta = 0$. As we will see the rate at which $\lambda(n, k) \to 0$ as $n \to \infty$ depends upon the nature of the minimum of $c(e^{i\theta})$. Let us *assume* that

$$\begin{aligned} &\text{i. } c(e^{i\theta}) > 0 && \theta \neq 0; \\ &\text{ii. } c(e^{i\theta}) \sim \sigma|\theta|^{\alpha} && \theta \to 0; \end{aligned} \tag{4.2.1}$$

where $\sigma > 0$, $\alpha > 0$.

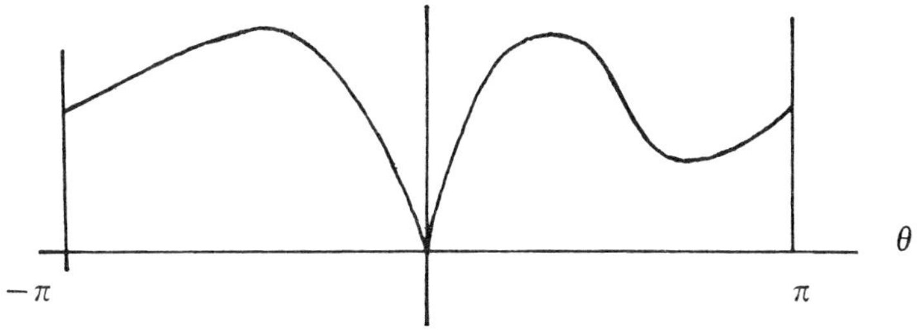

Fig. 4.1.

In order to see what to expect we proceed as follows. Let $\varepsilon > 0$ be fixed. By Corollary 1b, if $\{\theta | c(e^{i\theta}) = \varepsilon\}$ has measure 0, then $\lim_{n\to\infty} (n+1)^{-1} N(n; 0, \varepsilon) = (2\pi)^{-1} M(0, \varepsilon)$. A simple computation shows that if (1) holds then $M(0, \varepsilon) \sim 2(\varepsilon/\sigma)^{1/\alpha}$ as $\varepsilon \to 0+$, and if $\varepsilon = \lambda(n, k)$ then $(n+1)^{-1} N(n; 0, \varepsilon) = k/n+1$, from which it follows (formally) that $\lambda(n, k) \sim \sigma(\pi k/n+1)^{\alpha}$ as $n \to \infty$. The above derivation is justifiably suspect since we have permitted ε to vary with n; however, it makes plausible the following result.

Theorem 2a. Let c be real and continuous and satisfy conditions (4.2.1). Then for $k = 1, 2, \ldots$ fixed we have

$$\lambda(n, k) \sim \mu(k)/(n+1)^{\alpha} \qquad \text{as} \quad n \to \infty \tag{4.2.2}$$

where $0 < \mu(1) \leqq \mu(2) \leqq \mu(3) \leqq \ldots, \lim_{k\to\infty} \mu(k) = \infty$, are constants depending only upon σ and α.

We will not *prove* Theorem 2a but we will describe *how* to prove it. While it is not true except when $\alpha = 2$ that $\mu(k) = \sigma(\pi k)^{\alpha}$ $k = 1, 2, \ldots$, it can be shown that, interestingly enough,

$$\mu(k) \sim \sigma(\pi k)^{\alpha} \qquad k \to \infty$$

so that the informal argument preceding Theorem 2a gives an 'almost correct' result.

Problems of this kind were first studied by Kac, Murdock, and Szegö [40] who showed by an elementary argument that in the case $\alpha = 2$, (4.2.2) holds with $\mu(k) = \sigma(\pi k)^2$. The general solution which requires quite sophisticated methods is due to Parter [53], [54], [55], and Widom [73], [74], [75]. The definitive version of these researches is that given by Widom in [75].

We now proceed to sketch the proof of Theorem 2a. For $g \in \mathscr{L}^2(T)$,

$$g(e^{i\theta}) \sim \sum_{k=-\infty}^{\infty} g(k)e^{ik\theta},$$

we set

$$E(n)g \cdot (e^{i\theta}) = E^+(0)E^-(n)g \cdot (e^{i\theta}) = \sum_{k=0}^{n} g(k)e^{ik\theta},$$

and

$$Vg \cdot (e^{i\theta}) = c(e^{i\theta})g(e^{i\theta}).$$

$E(n)$ is a projection and V a bounded self-adjoint operator on $\mathscr{L}^2(T)$. It is apparent that $W^+[c, n]$ is unitarily equivalent to

$$E(n)VE(n)\Big|_{E(n)\mathscr{L}^2(T)},$$

the restriction of $E(n)VE(n)$ to the subspace $E(n)\mathscr{L}^2(T)$.

Let $\mathscr{L}^2(E)$ be the Hilbert space of complex functions on the line determined by the inner product

$$(f|g) = (2\pi)^{-1}\int_{-\infty}^{\infty} f(u)g(u)^* \, du.$$

For each $n = 1, 2, \ldots$ we define a mapping $\chi(n)$ from $\mathscr{L}^2(T)$ to $\mathscr{L}^2(E)$ by the formula

$$\chi(n)f \cdot (u) = \begin{cases} n^{-\frac{1}{2}}f(e^{iu/n}) & |u| \leqq n\pi \\ 0 & |u| > n\pi. \end{cases}$$

$\chi(n)$ is an isometric mapping of $\mathscr{L}^2(T)$ into $\mathscr{L}^2(E)$. The adjoint $\chi(n)^*$ of $\chi(n)$, given by

$$\chi(n)^*f \cdot (e^{i\theta}) = n^{\frac{1}{2}}f(n\theta) \qquad -\pi \leqq \theta \leqq \pi,$$

is a partially isometric mapping of $\mathscr{L}^2(E)$ onto $\mathscr{L}^2(T)$.

We now *transfer* $E(n)$ and V to $\mathscr{L}^2(E)$ by setting

$$F(n) = \chi(n)E(n)\chi(n)^*,$$
$$V(n) = \chi(n)V\chi(n)^*.$$

The operator

$$F(n)V(n)F(n)\Big|_{F(n)\mathscr{L}^2(E)}$$

is unitarily equivalent to

$$E(n)VE(n)\Big|_{E(n)\mathscr{L}^2(T)}\cdot$$

Thus if $U(n) = (n+1)^\alpha V(n)$ the eigen values of

$$F(n)U(n)F(n)\Big|_{F(n)\mathscr{L}^2(E)}$$

are $\{(n+1)^\alpha \lambda(n, k)\}_{k=1}^{n+1}$.

Let us set

$$s(n, u) = \begin{cases} (n+1)^\alpha c(e^{iu/n}) & |u| \leqq n\pi \\ 0 & |u| > n\pi, \end{cases}$$

$$s(u) = \sigma|u|^\alpha \qquad -\infty < u < \infty.$$

Then for $g \in \mathscr{L}^2(E)$

$$U(n)g \, . \, (u) = s(n, u)g(u).$$

Let us define

$$Ug \, . \, (u) = s(u)g(u)$$

for all g such that $s(u)g(u) \in \mathscr{L}^2(E)$. Then U is a non-negative self adjoint operator.

Since $U(n) \geqq 0$ and $U \geqq 0$ these operators have unique non-negative square roots.

Lemma 2b. Under assumptions i. and ii. of (1) $U(n)^{\frac{1}{2}}$ converges strongly to $U^{\frac{1}{2}}$ as $n \to \infty$.

Proof. We have

$$U^{\frac{1}{2}}f \cdot (u) = s(u)^{\frac{1}{2}}f(u),$$

$$U(n)^{\frac{1}{2}}f \cdot (u) = s(n, u)^{\frac{1}{2}}f(u).$$

Conditions i. and ii. of (4.2.1) imply that

$$\lim_{n\to\infty} s(n, u) = s(u) \qquad -\infty < u < \infty,$$

and that

$$0 \leqq s(n, u) \leqq As(u) \qquad -\infty < u < \infty, n = 1, 2, \ldots$$

for a suitable constant A. We may therefore apply the Lebesgue limit theorem to show that if

$$\int_{-\infty}^{\infty} |s(u)^{\frac{1}{2}} f(u)|^2 \, du < \infty,$$

then

$$\int_{-\infty}^{\infty} |\{s(u)^{\frac{1}{2}} - s(n, u)^{\frac{1}{2}}\} f(u)|^2 \, du \to 0 \quad \text{as} \quad n \to \infty.$$

The same argument shows that U is the strong limit of $U(n)$ as $n \to \infty$. While this might seem more natural, the result we have proved is what would be needed if we were to carry out the proof of Theorem 2a to the end.

For $g \in \mathscr{L}^2(E)$,

$$g(u) = \int_{-\infty}^{\infty} g(z) e^{iuz} \, dz,$$

we set

$$Fg \cdot (u) = \int_{-1}^{1} g(z) e^{iuz} \, dz.$$

F is a projection on $\mathscr{L}^2(E)$.

Lemma 2c. F is the strong limit of $F(n)$ as $n \to \infty$.

Proof. Let $f(u)$ be continuous and such that $f(u) = 0(|u|^{-2})$ as $u \to \pm\infty$. We have

$$F(n)f = \chi(n)E(n)\chi(n)^* f.$$

Now

$$\chi(n)^* f \cdot (\theta) = n^{\frac{1}{2}} f(n\theta) \qquad \theta \in T,$$

$$E(n)\chi(n)^* f \cdot (\theta) = \sum_{k=-n}^{n} e^{ik\theta} (2\pi)^{-1} \int_{-\pi}^{\pi} n^{\frac{1}{2}} f(n\varphi) e^{-ik\varphi} \, d\varphi,$$

$$F(n)f(u) = n^{-1} \sum_{k=-n}^{n} e^{iku/n} (2\pi)^{-1} \int_{-n\pi}^{n\pi} f(v) e^{-ikv/n} \, dv \tag{4.2.3}$$

$$= n^{-1} \sum_{k=-n}^{n} f\left(\frac{k}{n}\right) e^{iku/n} + \varepsilon(n, u)$$

where $\varepsilon(n, u) \to 0$ as $n \to \infty$ uniformly for u in any bounded set.

Here

$$\hat{f}(z) = (2\pi)^{-1} \int_{-\infty}^{\infty} f(u) e^{-iuz} du.$$

It follows that

$$\lim_{n\to\infty} F(n)f \cdot (u) = \int_{-1}^{1} \hat{f}(z) e^{iuz} du = Ff \cdot (u)$$

uniformly for u in any bounded set. If $g(u)$ has bounded support, this implies that

$$\lim (F(n)f \,|\, g) = (Ff \,|\, g). \tag{4.2.4}$$

Since $||F(n)|| = 1$, $n = 1, \ldots$ it follows that (4.2.4) holds for all $f, g \in \mathscr{L}^2(E)$. Thus $F(n)$ converges weakly to F; but for projections weak convergence implies strong convergence.

Lemmas 2b and 2c suggest that $F(n)U(n)F(n)$ restricted to $F(n)\mathscr{L}^2(E)$ converges to FUF restricted to $F\mathscr{L}^2(E)$, and therefore that

$$\lim_{n\to\infty} (n+1)^{\alpha} \lambda(n, k) = \mu(k) \qquad k = 1, 2, \ldots,$$

where $\{\mu(k)\}_{k=1}^{\infty}$ are the eigen values of FUF restricted to $F\mathscr{L}^2(E)$ written in non-decreasing order. Properly interpreted this turns out to be true although it is far from trivial. For the details of these arguments we refer the reader to Widom's paper [75], and also to the author's paper [27] which contains a careful analysis of the perturbation theory involved.

In Theorem 2a we assumed that $c(e^{i\theta})$ had a unique minimum. The case where $c(e^{i\theta})$ takes on its minimum value at several points has been treated by Kesten [44].

These methods can also be applied to Toeplitz operators constructed using orthogonal polynomials $\{p_k(x)\}_0^{\infty}$ in place of $\{e^{ik\theta}\}_{-\infty}^{\infty}$, etc. See the papers of Hirschman [27] and Davis [15].

4.3. A SECOND ORDER APPROXIMATION

Let us assume that the minimum of $c(e^{i\theta})$ (which we assume to be equal to 0 and to be taken at $\theta = 0$) is 'very regular'. Specifically we suppose that:

i. $c(e^{i\theta})$ is real and continuous for $\theta \in T$, $c(e^{i\theta})|_{\theta=0} = 0$

and $c(e^{i\theta}) > 0$ for $0 < |\theta| \leqq \pi$; (4.3.1)

ii. $c(e^{i\theta})$ is four times continuously differentiable in a neighborhood of $\theta = 0$ and

$$\frac{d}{d\theta} c|_{\theta=0} = 0, \qquad \left(\frac{d}{d\theta}\right)^2 c|_{\theta=0} = \tau^2 > 0.$$

We know from 4.2 that in this case

$$\lambda(n, k) = \frac{\tau^2\pi^2k^2}{2(n+1)^2} + o(n^{-2}) \quad \text{as} \quad n \to \infty \tag{4.3.2}$$

for each $k = 1, 2, \ldots$. In a remarkable paper written in 1958, [71], Widom established the following more precise result.

Theorem 3a. Let i and ii hold and let

$$\alpha = \alpha(c) = \frac{1}{2\pi}\int_T \log\left[\frac{c(e^{i\theta})}{2\tau^2}\cot^2\frac{\theta}{2}\right]\csc^2\theta\, d\theta;$$

then for each $k = 1, 2, \ldots$ we have

$$\lambda(n, k) = \frac{\tau^2\pi^2k^2}{2(n+1)^2}\left[1 + \frac{\alpha}{n+1}\right] + o(n^{-3}) \quad \text{as} \quad n \to \infty. \tag{4.3.3}$$

A point of interest in (4.3.3) is that the coefficient α depends upon the global behaviour of c and not just on the behaviour of c at $\theta = 0$.

The complete demonstration of Theorem 3a is too long to include. We will, however, explain the idea on which this demonstration is based. Let

$$c(z) = \sum_{-j}^{h} c(r)z^r \tag{4.3.4}$$

where $j, h > 0$ and where $c(-j), c(h) \neq 0$ and let $\rho_1(\lambda), \ldots, \rho_{j+h}(\lambda)$ be the roots of the equation

$$z^j\{c(z) - \lambda\} = 0. \tag{4.3.5}$$

The following identity, in a slightly less precise form, was given by Widom in [71]. In the present form it is due to Baxter and Schmidt [6].

Theorem 3b. Under the above assumptions $D[n, c-\lambda]$ is equal to

$$(-1)^{(n+1+h)j} c(j)^{n+1} \sum_{I} \Big(\prod_{\nu \in I} \rho_\nu\Big)^{-n-1} \Big(\prod_{\mu \notin I} \rho_\nu\Big)^{j} \prod_{\substack{\nu \in I \\ \mu \notin I}} (\rho_\nu - \rho_\mu)^{-1}, \qquad (4.3.6)$$

where I runs over all subsets of cardinality j of the set $\{1, 2, \ldots, h+j\}$.

Widom's proof of Theorem 3a now proceeds as follows. Suppose that $c(e^{i\theta})$ is a trigonometric polynomial satisfying the assumptions (4.3.1). A fairly intricate computation based upon Theorem 3b shows that (4.3.3) holds. The general case is then obtained by an approximation argument using the Weyl-Courant lemma.

This method can be extended to apply to other more general Toeplitz operators. See Hirschman [26], Davis [14], and Parter [57].

4.4. Variable coefficient Toeplitz matrices

In the application of difference methods to the solution of partial differential equations with variable coefficients matrices of the form

$$\left[c\left(\frac{j}{n}, \frac{k}{n}, j-k\right)\right]$$

arise where $c(x, y, k)$ is a complex function defined for $-\infty < x < \infty$, $-\infty < y < \infty$, $k = \ldots, -1, 0, 1, \ldots$. Some of the results we have discussed can be extended to this case. In particular Kac, Murdock, and Szegö in [40] obtained an analogue of Theorem 1a for the variable coefficient case. More recently Parter [56] has found an asymptotic formula for extreme eigen values of certain variable coefficient Toeplitz matrices, which is closely related to the material of 4.2. See also Hirschman [28], and Davis and Hirschman [16].

References

[1] Achiezer, N. I., 'The continuous analogues of polynomials orthogonal on systems of intervals', *Doklady Akad. Nauk*, S.S.S.R., 141 (1961), pp. 263–266.

[2] Achiezer, N. I., 'The continuous analogues of polynomials orthogonal on a circle', *Doklady Akad. Nauk*, S.S.S.R., 141 (1961), pp. 769–772.

[3] Achiezer, N. I., 'The continuous analogues of some theorems on Toeplitz matrices', *Ukrainian Math. J.*, 16 (1964), pp. 445–462.

[4] Arens, R. and Calderon, A., 'Analytic functions of several Banach algebra elements', *Ann. of Math.*, 62 (1955), pp. 204–216.

[5] Baxter, Glen., 'Polynomials defined by a difference system', *J. Math. Anal. and Appl.*, 2 (1961), pp. 223–263.

[6] Baxter, Glen., 'A convergence equivalence related to polynomials on the unit circle', *Trans. Amer. Math. Soc.*, 99 (1961), pp. 471–487.

[7] Baxter, Glen., 'An asymptotic result for the finite predictor', *Math. Scandinavica*, 10 (1962), pp. 137–144.

[8] Baxter, Glen., 'A norm inequality for a 'finite section' Wiener-Hopf equation', *Illinois J. Math.*, 7 (1963), pp. 97–103.

[9] Baxter, Glen., and P. Schmidt, 'Determinants of a certain class of non-Hermitian Toeplitz matrices', *Math. Scandinavica*, 9 (1961), pp. 122–128.

[10] Baxter, Glen., and Hirschman, Jr. I. I., 'An explicit inversion formula for finite section Wiener-Hopf operators', *Bull. Amer. Math. Soc.*, 70 (1964), pp. 820–823.

[11] Bowers, Newton L., Jr., '*Toeplitz Forms Associated with Matrix Valued Functions*', Thesis, Univ. of Minnesota, 1965.

[12] Calderon, A., Spitzer, F., and Widom, H., 'Inversion of Toeplitz matrices', *Illinois J. Math.*, 3 (1959), pp. 490–498.

[13] Courant, R. and Hilbert, D., *Methods of Mathematical Physics*, Vol. I, New York, 1953.

[14] Davis, Jeffrey, 'Extreme eigen values of Toeplitz operators of the Hankel type, I', *J. Math. and Mech.*, 14 (1965), pp. 245–276.

[15] Davis, Jeffrey, 'On the extreme eigen values of Toeplitz operators of the Hankel type, II', *Trans. Amer. Math. Soc.*, 116 (1965), pp. 267–299.

[16] Davis, Jeffrey, and Hirschman, Jr., I. I., 'Toeplitz forms and ultraspherical polynomials', in *Pacific J. Math.*, 18 (1966), pp. 73–95.

[17] Devinatz, A., 'Toeplitz operators on H^2 spaces', *Trans. Amer. Math. Soc.*, 112 (1964), pp. 304–317.

[18] Devinatz, A., 'Asymptotic estimates for the finite predictor', *Math. Scand.*, 15 (1964), pp. 111–120.

[19] Devinatz, A., 'Conjugate function theorems for Dirichlet algebras', to appear in *Rev. Un. Mat. Argentina.*

[20] Devinatz, A., 'An extension of a limit theorem of G. Szegö', *J. of Math. Anal. and Appl.*, 14 (1966), pp. 499–510.

[21] Devinatz, A., 'The strong Szegö limit theorem', *Illinois J. of Math.*, 11 (1967), pp. 160–175.

[22] Devinatz, A., 'On Wiener-Hopf operators', *Functional Analysis.* Proceedings of a Conference held at the University of California, Irvine (1967), pp. 81–118.

[23] Geronimus, L. Ya., *Orthogonal Polynomials.* New York, 1961.

[24] Grenander, U. and Szegö, G., *Toeplitz Forms and their Applications.* Univ. of California Press, 1958.

[25] Hille, E. and Phillips, R., 'Functional Analysis and Semi-Groups', *Amer. Math. Soc. Colloquium Publications*, 31 (1957).

[26] Hirschman, I. I., Jr., 'Extreme eigen values of Toeplitz forms associated with orthogonal polynomials', *J. d'Anal. Math.*, 12 (1964), pp. 187–242.

[27] Hirschman, I. I., Jr., 'Extreme eigen values of Toeplitz forms associated with Jacobi polynomials', *Pacific J. Math.*, 14 (1964), pp. 107–161.

[28] Hirschman, I. I., Jr., 'Extreme eigen values of Toeplitz forms associated with ultraspherical polynomials', *J. of Math. and Mech.* 13 (1964), No. 2, pp. 249–282.

[29] Hirschman, I. I., Jr., 'Finite sections of Wiener-Hopf equations and Szegö polynomials', *J. Math. Anal. and Appl.*, 11 (1965), pp. 290–320.

[30] Hirschman, I. I., Jr., 'On a theorem of Szegö, Kac, and Baxter', *J. d'Analyse Math.*, 14 (1965), pp. 225–234.

[31] Hirschman, I. I., Jr., 'Finite section Wiener-Hopf equations on a compact group with ordered dual', *Bull. Amer. Math. Soc.*, 70 (1964), pp. 508–510.

[32] Hirschman, I. I., Jr., 'The strong Szegö limit theorem for Toeplitz determinants', *Amer. Jour. Math.*, 88 (1966), pp. 577–614.

[33] Hirschman, I. I. ,Jr., 'Szegö polynomials on a compact group with ordered dual', *Canadian J. Math.*, 18 (1966), pp. 538–560.

[34] Hirschman, I. I., Jr., 'Szegö polynomials on a locally compact Abelian group with ordered dual', *Trans. Amer. Math. Soc.*, 121 (1966), pp. 133–159.

[35] Hirschman, I. I., Jr., 'On a formula of Kac and Achiezer', *J. Math. and Mech.*, 16 (1966), pp. 167–196.

[36] Hirschman, I. I., Jr., 'The spectra of certain Toeplitz matrices', *Illinois J. Math.*, 11 (1967) pp. 145–159.

[37] Kac, M., 'Toeplitz matrices, translation kernels, and a related problem in probability theory', *Duke Math. J.*, 21 (1954), pp. 501–509.

[38] Kac, M., *Probability and Related Topics in Physical Sciences*. New York, 1959.

[39] Kac, M., 'Some combinatorial aspects of the theory of Toeplitz matrices', to appear in the *Proceedings of the IBM Conference on Combinatorial Mathematics*.

[40] Kac, M., Murdock, W., and Szegö, G., 'On the eigen values of certain Hermitian forms', *J. Mat. Mech. Anal.*, 2 (1953), pp. 767–800.

[41] Kalafati, P., 'On a new system of orthogonal functions', *Doklady Akad. Nauk*, S.S.S.R.' 105 (1955), pp. 631–633.

[42] Kesten, H., 'On a theorem of Spitzer and Stone and random walks with absorbing barriers', *Ill. J. Math.*, 5 (1961), pp. 246–266.

[43] Kesten, H., 'Random walks with absorbing barriers and Toeplitz forms', *Ill. J. Math.*, 5 (1961), pp. 267–290.

[44] Kesten, H., 'On the extreme eigenvalues of translation kernels and Toeplitz matrices', *J. d'Analyse Math.*, 10 (1962–63), pp. 117–138.

[45] Kolmogoroff, A. N., 'Stationary sequences in Hilbert space', *Bull. Univ. Moscow*, 2 (1941), 40 pages.

[46] Krein, M. G., 'On a generalization of some investigations of G. Szegö, W. M. Smirnov, and A. N. Kolmogoroff', *Doklady Akad. Nauk*, S.S.S.R., 46 (1945), pp. 91–94.

[47] Krein, M. G., 'On integral equations generating differential equations of second order', *Doklady Akad. Nauk*, S.S.S.R., 97 (1954), pp. 21–24.

[48] Krein, M. G., 'On the determination of a particle potential by its *S*-function', *Doklady Akad. Nauk*, S.S.S.R., 105 (1955), pp. 433–436.

[49] Krein, M. G., 'The continuous analogues of theorems on polynomials orthogonal on the unit circle', *Doklady Akad. Nauk,* S.S.S.R., 104 (1955), pp. 637–640.

[50] Krieger, H. A., 'Toeplitz operators on locally compact Abelian groups', *J. Math. and Mech.*, 14 (1965), pp. 439–478.

[51] Lenard, A. 'Momentum distribution in the ground state of the one-dimensional system of impenetrable Bosons' *Jour. of Math. Physics* 5 (1964), pp. 930–943.

[52] Mejlbo, L. and Schmidt P., 'On the eigen values of generalized Toeplitz matrices', *Math. Scand.*, 10 (1962), pp. 5–16.

[53] Parter, S. V., 'Extreme eigen values of Toeplitz forms and applications to elliptic differential equations', *Trans. Amer. Math. Soc.*, 99 (1961), pp. 153–193.

[54] Parter, S. V., 'On the extreme eigen values of truncated Toeplitz matrices' *Bull. Amer. Math. Soc.*, 67 (1961) pp. 191–196.

[55] Parter, S. V., 'On the extreme eigen values of Toeplitz matrices', *Trans. Amer. Math. Soc.*, 100 (1961), pp. 263–276.

[56] Parter, S. V., 'On eigen values of certain generalizations of Toeplitz matrices'. *Arch. Rat. Mech. and Anal.* 11 (1962), pp. 244–257.

[57] Parter, S. V., 'Remarks on the extreme eigen values of Toeplitz forms associated with orthogonal polynomials' *J. Math. and Appl.* 12 (1965) pp. 456-470.

[58] Reich, E. 'On non-Hermitian Toeplitz matrices' *Math. Scand.*, 10 (1962) pp. 145–152.

[59] Schmidt, P. and Spitzer F., 'The Toeplitz matrices of an arbitrary Laurent polynomial', *Math. Scandinavica*, 8 (1960), pp. 15–38.

[60] Shinbrot, M., 'A class of difference kernels', *Proc. Amer. Math. Soc.*, 13 (1962), pp. 399–406.

[61] Smithies, F., 'The Fredholm theory of integral equations', *Duke Math. J.*, 8 (1941) pp. 107–130.

[62] Spitzer, F. L., *Principles of Random Walk.* Van Nostrand, Princeton, N. J., 1964.

[63] Spitzer F. L., and Stone C. J., 'A class of Toeplitz forms and their application to probability theory' *Ill. J. Math.*, 4 (1960), pp. 253–277.

[64] Szegö, G., 'Ein Grenszwertatz uber die Toeplitzschen Determinanten einer reellen positiven Funktion', *Math. Annalen,* 76 (1915), pp. 490–503.

[65] Szegö, G., 'Orthogonal polynomials', *Amer. Math. Soc. Colloquium Publ.*, 23, 1939.

[66] Szegö, G., 'On certain Hermitian forms associated with the Fourier series of a positive function', Festkrift Marcel Riesz, Lund (1952), pp. 228–238. (*Commun. du Seminaire Math. de l'Univ. de Lund*).

[67] Titchmarsh, E. C., *The Theory of Functions.* Oxford, 1939.

[68] van der Waerden, B. L., *Moderne Algebra.* New York, 1943.

[69] Walsh, J. L., 'Interpolation and Approximation by Rational Functions the Complex Domain', *Colloquium Publications of the Amer. Math. Soc.*, 20, rev. ed., 1956.

[70] Watanabe, T., 'On a theorem of F. L. Spitzer and C. J. Stone', *Proc. Japan. Academy,* 37 (1961), pp. 346–351.

[71] Widom, H., 'On the eigen values of certain Hermitian operators', *Trans. Amer. Math. Soc.*, 88 (1958), pp. 491–522.

[72] Widom, H., 'A theorem on translation kernels in *n*-dimensions', *Trans.*

Amer. Math. Soc., 94 (1960), pp. 170–180.

[73] Widom, H., 'Extreme eigen values of translation kernels', *Trans. Amer. Math. Soc.*, 100 (1961), pp. 252–262.

[74] Widom, H., 'Stable processes and integral equations', *Trans. Amer. Math. Soc.*, 98 (1961), pp. 430–439.

[75] Widom, H., 'Extreme eigen values of N-dimensional convolution operators', *Trans. Amer. Math. Soc.*, 106 (1963), pp. 391–414.

Matching theory, an introduction

by Gian-Carlo Rota

MASSACHUSETTS INSTITUTE OF TECHNOLOGY

CAMBRIDGE, MASSACHUSETTS

and L. H. Harper

THE ROCKEFELLER UNIVERSITY

NEW YORK, NEW YORK

1. Introduction

Given a chessboard of size n by n. With a pair of scissors, we clip out arbitrarily a number of squares. We are to cover every square of the amputated chessboard by non-overlapping domino pieces, each piece covering exactly two adjacent squares, either vertically or horizontally. Under what conditions can the amputated chessboard be so covered? If so, how many distinct coverings are there? If no covering exists, what is the maximum number of squares we can cover?

This is a typical problem of matching theory. As stated, it is by no means a difficult one, but, frivolous as it may sound, it did arise in an important context in statistical mechanics. Physicists call the dominoes 'dimers' and refer to the covering as a 'dimer covering'.

Now color each square of the chessboard either white or black, so that adjacent squares are assigned different colors. A domino piece placed on the chessboard will cover two squares of different colors. Therefore, a *necessary* condition that the covering be possible, is that white and black squares be present in equal numbers on the amputated chessboard. We shall see that this necessary condition can be strengthened without difficulty to a necessary and sufficient condition for a covering to exist, for a board of arbitrary shape. The resulting theorem, commonly called the *Marriage Theorem*, applies to a variety of similar and more intricate situations.

Roughly speaking, matching theory is concerned with the possibility and the number of ways of covering a large, irregularly shaped combinatorial object with several replicas of a given small, regularly shaped object, subject usually to the requirement that the small objects shall not overlap and that the small objects be 'lined up' in some sense or other; for example, in the domino problem this 'lining up' requirement corresponds to the fact that each domino must cover two squares of different colors. Variants of this main problem relax the requirement that the small objects shall not overlap, and replace it with certain prescribed forms of overlapping; for example, Sperner's Theorem (v. below) gives the maximum size of a family of subsets of a finite set, subject to the condition that no one set in the family shall contain another; there is a similar result for subspaces of a finite-dimensional vector space over a Galois field.

In this paper we present results of matching theory that we believe to be both elegant and important enough to encourage the reader to pursue the matter further by himself. It is commonly thought that the consequences of combinatorial theorems are limited and must in fact be combinatorial. We have selected from a vast amount of available material those results which best show disparate branches of mathematics relating to a common combinatorial theme: matching. Thus, after presenting Dilworth's Theorem, the Marriage Theorem, and a few other basic results, we digress into the representation of families of sets by n-tuples of integers, a problem which is closely related to current ideas on 'complexity' of mathematical structures. The proof of the Coding Theorem gives an occasion for simultaneously developing the representation theory of distributive lattices, a topic that has been strangely neglected in favor of the less interesting Boolean algebras. The Marriage Theorem is then applied to derive the Birkhoff-von Neumann Theorem, and this in turn leads to a digression into inequalities for symmetric polynomials in several variables.

In Section 8 a measure-theoretic extension of matching is developed, giving the solution of Birkhoff's Problem 111. This provides a continuous version of matching, something often sought after, but seldom successfully established. This in turn provides one of the best confirmations of our thesis that combinatorial theory is not a hide-bound subject. In the same spirit Sperner's Theorem is extended (Section 9) and applied to solve a problem first posed by Littlewood and Offord in connection with some estimates on the zeroes of random polynomials (Section 10). Finally, the last section (Section 11) demonstrates the existence of the mean for almost periodic functions as an application of the Marriage Theorem.

Little of the present material has been previously collected. In the case of the better-known matching theorems, we have tried to present either different proofs, or different connections with the rest of the theory. Although the present subject is nominally identical with that of the recent survey by Mirsky and Perfect, the actual material presented is almost disjoint and the point of view orthogonal. We hope the two works will complement each other.

As an enticement to the reader, the final section has been devoted to historical comments and especially to the statements of research problems. These range from the most difficult and important problems of combinatorial theory to curiosities that happen to connect with the text. We have made no effort to separate the important from the peripheral, nor the easy from the difficult; our hope is that these problems will add to the reader's understanding of the text and stimulate research along the indicated lines.

One major topic has been neglected throughout, namely, the connections between matching theory and linear algebra. It is not unlikely that this entire subject will some day become an application of the methods of exterior algebra. We defer the discussion to later publications, the present one being meant as a first installment.

2. Basic Definitions

We omit proofs of the statements in this Section – they can be easily supplied, or looked up in a number of standard texts. We also omit any mention of the word 'finite': it is understood that all objects described below are finite, unless the contrary is explicitly stated. The reader is assumed to be familiar with the rudiments of set theory and linear algebra.

After the notions of set and function, the most fundamental is that of a *partition* of a set S. A partition is a family of non-empty subsets of S, called *blocks*, mutually disjoint, which cover the set S. The partition consisting of only one block, namely, the set S itself, is denoted by I, whereas the partition where every element is a block is written 0. A *set of representatives* of a partition π of a set S, is a subset of S which meets every block of π in exactly one element. A set of common representatives of two partitions of the same set is a subset which meets each block of either partition in exactly one element.

The next fundamental notion is that of a *simplicial complex* defined on a set S. A simplicial complex is a family $\boldsymbol{C}$ of subsets of S, with the property that if $A \in \boldsymbol{C}$ and $B \subseteq A$, then $B \in \boldsymbol{C}$. The elements of $\boldsymbol{C}$ are usually called *faces*. Simplicial complexes have been studied mostly in topology, where they received their name, but they occur just as often in combinatorics, sometimes under hard-to-recognize guises (for example, in the theory of permutations with restricted position).

A face containing $n+1$ elements in a simplicial complex is sometimes said to be of *dimension n*.

The most intensively studied among simplicial complexes are the *linear graphs*. These are simplicial complexes in which every face is of dimension zero or one. The faces of dimension zero are called *vertices*, whereas those of dimension one are called *edges*. Needless to say, one pictures the edges of a graph as straight line segments joined together at the vertices (it is easily shown that every graph can be imbedded in three space using straight-line segments only).

An *oriented* linear graph is a linear graph where each edge is an ordered pair of vertices, called the *head* and *tail* and pictured as an arrow. A *general* graph (sometimes called a multigraph) consists of a set V, whose elements are called vertices, another set E, whose elements are called edges, together with two functions t (for tail) and h (for head) from E to V. The geometric interpretation is immediate: several edges are 'allowed' to have the same head and tail. An edge whose head and tail coincide is a *loop*.

A *relation* between two sets A and B – called the domain and the range – is a subset of the product $A \times B$ of the sets. To every relation one associates a graph, called a *bipartite graph*, whose vertices are the elements of $A \cup B$, and whose edges are the ordered pairs (a, b) belonging to the relation R.

An *ordered set* (also known as a *partially* ordered set) P is a set together with a binary relation $x \leqq y$, defined for some pairs x, y ranging over P, which is reflexive, antisymmetric and transitive. Commonplace though this notion is nowadays, we repeat the definition in full. The defining properties are:

$$x \leqq x \tag{2.1}$$

$$\text{If } x \leqq y \text{ and } y \leqq x, \text{ then } x = y \tag{2.2}$$

$$\text{If } x \leqq y \text{ and } y \leqq z, \text{ then } x \leqq z. \tag{2.3}$$

Two elements x and y are said to be *comparable* when $x \leqq y$ or $y \leqq x$; otherwise, they are said to be *incomparable* or *unrelated*. One writes $x > y$ when $x \geqq y$ but $x \neq y$; we shall sometimes use the expression 'x lies above y' to mean $x > y$.

We proceed to get some of the standard jargon of ordered set theory out of the way. A *maximal* element m in an ordered set P is one with the property that, if $x \geqq m$, then $x = m$. A *minimal* element is similarly defined. If an ordered set has a unique maximal element, it is denoted by I; similarly, a unique minimal element is denoted by 0. A *segment* $[x, y]$ in an ordered set P consists of all elements z of P such that $x \leqq z \leqq y$, with the induced ordering. An ordered set is *linearly ordered* when every pair of elements is comparable. An ordered *tree* is an ordered set in which every segment is linearly ordered. An ordered tree having a unique minimal element 0 is said to be a *rooted tree*, and 0 is then called the *root*.

In an ordered set, an element x *covers* an element y when $x > y$ and if $x \geqq z \geqq y$, then $z = x$ or $z = y$. An ordered set is *locally finite* when every segment is finite; clearly every finite ordered set is locally finite. With every locally finite ordered set P one can associate an oriented linear graph H, called the *Hasse diagram* of P, and defined as follows. The vertices of H are a replica of the elements of P (and we shall identify them with the elements of P); two vertices x and y of H are joined by an edge whenever x covers y; the vertex x is the tail and the vertex y is the head of the edge.

The *dual* P^* of an ordered set P is defined as follows: The elements of P^*, considered as a set, are identified with the elements of P (that is, there is a canonical one-to-one map of P^* as a set onto P as a set), and the order relation on P^* is defined as follows: $x \leqq y$ in P^* whenever $x \geqq y$ in P. Note that we use the same symbol $\leqq$ for both the order relation of P and that of P^*; we shall often do this, and it will cause little confusion. If P is locally finite, the Hasse diagram of P^* is obtained from the Hasse diagram of P by reversing the orientation of every edge.

An *atom* in an ordered set having a unique minimal element 0 is an element a of P which covers 0. An element c of P which is covered by I – whenever I exists – is similarly called a coatom or dual atom. Sometimes an atom is also called a *point* (by Dilworth, for instance) but we shall try not to.

A *lattice* is an ordered set P with the property that, given any two elements x and y of P, the subset $\{z: z \geqq x \text{ and } z \geqq y\}$ of P, with the induced ordering, has a unique minimal element, denoted by $x \vee y$ and called the *least upper bound* (sup) or *join* of x and y; and furthermore, the subset $\{u: u \leqq x \text{ and } u \leqq y\}$ has a unique maximal element, denoted by $x \wedge y$ and called the *greatest lower bound* (inf) or *meet* of x and y. It follows immediately from the definition that each of the operations $\vee$ and $\wedge$ is commutative, associative and idempotent ($x \vee x = x$ and $x \wedge x = x$), and that together they satisfy the *absorption laws*

$$x \vee (y \wedge x) = x$$

$$x \wedge (y \vee x) = x$$

Conversely, if P is a set with two operations $\vee$ and $\wedge$ obeying the commutative, associative, idempotent and absorption laws, then one can define $x \geqq y$ as $x \vee y = x$ and this turns out to be an order relation. Relative to this order relation, P turns out to be a lattice, where joins and meets coincide with the originally given operations $\vee$ and $\wedge$.

In lattices arising from combinatorial objects, one of the two operations – usually the meet – is easier to define than the other. One can get around this difficulty by the following remark: any finite ordered set P in which meets are defined, and which has a unique maximal element I, is a lattice. Simply define the join $x \vee y$ as inf $\{z: z \geqq x \text{ and } z \geqq y\}$. The assumption that P has a unique maximal element insures that this set is not empty, and hence that the inf always exists if P is finite.

If S is a subset of a lattice L, we write $\vee S$ and $\wedge S$ to denote the join and meet of the elements of S.

The notion of isomorphism of two ordered sets is defined in the obvious way. An antiisomorphism of ordered sets P and Q is an isomorphism of P and Q^* (the dual of Q).

An isomorphism of one ordered set into another will also be called a *faithful representation*.

A lattice isomorphism of a lattice P into a lattice Q is a one-to-one map f of P into Q such that $f(x \vee y) = f(x) \vee f(y)$ and similarly $f(x \wedge y) = f(x) \wedge f(y)$. An isomorphism of ordered sets P into Q is not necessarily a lattice isomorphism.

A *chain* is a linearly ordered set. An ordered set satisfies the *chain condition* when it has a minimal element 0, and for every x in P all maximal chains with minimal element 0 and maximal element x have the same size. One then introduces the *rank* of an element x, written $r(x)$, as the size of a maximal chain between x and 0, *minus* 1. For example, the rank of an atom is 1, and $r(0) = 0$.

Three classes of lattices occur frequently:

2.1. The lattice of all subsets of a set S

We write $v(A)$ to denote the number of elements of a subset A, and call it the *size* (cardinality) of A; $v(A)$ is also the rank function in the lattice of subsets. The size of the family of all subsets of k elements out of a set of n elements is the *binomial coefficient* $\binom{n}{k}$, some of whose properties will be taken as known.

2.2. The lattice of subspaces

Consider an n-dimensional vector space V over a field F. Write $L(V)$ for the lattice of subspaces of V, ordered by inclusion. The meet of two subspaces R and S is their set-theoretic intersection, and their join is the vector subspace jointly spanned by R and S, sometimes written $R+S$. The lattice $L(V)$ satisfies the chain condition, where $r(S)$ is the dimension of subspace S, usually written dim (S).

The combinatorially interesting class of lattices occurs when the field F is finite, with q elements – q is a power of a prime. Then the lattice $L(V)$ is finite, and the number of subspaces of dimension k is the *Gaussian polynomial* or q-binomial coefficient

$$\begin{bmatrix} n \\ k \end{bmatrix}$$

Since the properties of Gaussian polynomials are less well-known than those of binomial coefficients, we now derive a specific formula for them.

If S is a k-dimensional vector space over the finite field F, we need a simple formula for the number of one-to-one linear transformations of S onto S. This is derived as follows.

Any independent set $x_1, x_2, \ldots, x_k$ of vectors in V determines a subspace of dimension k. The first vector x_1 can be chosen in q^n-1 ways, since there are q^n distinct vectors and we can choose any vector but the zero vector. The second vector x_2 can be chosen in q^n-q ways, since we can choose any vector not on the line spanned by x_1, and we can choose any vector not on this line. Proceeding in this way, we find that the set $\{x_1, x_2, \ldots, x_k\}$ of linearly independent vectors can be chosen in

$$(q^n-1)(q^n-q)(q^n-q^2)\ldots(q^n-q^{k-1})$$

ways. However, a subspace of dimension k has

$$(q^k-1)(q^k-q)\ldots(q^k-q^{k-1})$$

distinct bases, where a basis is considered as a linearly ordered set of vectors. The number of subspaces of dimension k equals therefore the ratio of these two expressions:

$$\begin{bmatrix} n \\ k \end{bmatrix} = \frac{(q^n-1)(q^n-q)\ldots(q^n-q^{k-1})}{(q^k-1)(q^k-q)\ldots(q^k-q^{k-1})}$$

$$= \frac{(q^n-1)(q^{n-1}-1)\ldots(q^{n-k+1}-1)}{(q^k-1)(q^{k-1}-1)\ldots(q-1)}.$$

2.3. Partitions of a set

If π and σ are partitions of a set S, set $\pi \leqq \sigma$ to mean that each block of π is contained in some block of σ. One says that π is a *refinement* of σ. The resulting ordered set of partitions, $\pi(S)$, is a lattice, where the meet $\pi \wedge \sigma$ is the partition whose blocks are obtained by intersecting each block of π with every block of σ, and discarding all empty blocks. The join $\pi \vee \sigma$ of two partitions is a little harder to describe and we shall use the device explained above. It is the 'finest' partition τ, subject to the condition that any two overlapping blocks of π and σ are subsets of the same block of τ. The partition 0 of a set S of size n has n blocks, whereas the partition I has one block. The number of blocks, minus one, is the rank function for the dual ordered set $\pi(S)$.* The number of partitions with k blocks in a set of size n is written $S(n, k)$ and called a *Stirling number of the second kind.*

It is important not to confuse the notion of partition of a set, just defined, with the related notion of *partition of a number n*, where n is a positive integer. A partition of a number n is a function α from the positive integers to the non-negative integers, with the property that

$$\sum_{k>0} k\alpha(k) = n.$$

The value $\alpha(k)$ of the function α is usually called 'the number of occurrences of the integer k'.

3. Partitions of an Ordered Set

The main results of matching theory have two peculiar properties: they are stated as minimax identities – or easily reduced to such – and they are all derivable from each other. This leaves us with a relatively free choice on where to start.

Dilworth's Theorem, which we prove below, is perhaps the most elegant matching theorem; in the opinion of Mirsky and Perfect, 'it is possible that Dilworth's Theorem is the most fundamental among the finite results.' Since its discovery in 1950, half a dozen proofs have been produced (the one we give is the simplest); yet, Dilworth's Theorem continues to exert a strange fascination. For these reasons we choose it as the starting point.

Given a finite ordered set P, we are to partition its elements into blocks, each of which shall be a linearly ordered set, or 'chain', relative to the ordering of P. This can be done in a trivial way, by taking the partition of P where each element is a block by itself. However, we add the requirement that the partition shall have a minimal number of blocks, and ask for this minimal number. The answer is given in terms of the notion of an *unrelated subset* of P; this is a subset U of P such that, if p and q are elements of U, then neither $p > q$ nor $p < q$. A partition of P into chains must have as many elements as there are in any of the unrelated subsets of P; it is the surprising conclusion of Dilworth's Theorem that this condition turns out to be sufficient as well:

Theorem 3.1. The minimum number of blocks in a partition of a finite ordered set into chains equals the maximum size of an unrelated subset.

The proof proceeds by induction on the size of P. The conclusion is trivial if P has one element. Assume that it is true for all ordered sets $Q \subsetneqq P$, with the relative ordering, and say the maximum size of an unrelated subset of P is n. There are two cases.

Case 1. There is an unrelated set U of size n, not containing all maximal nor all minimal elements of P.

Define P^+ as the subset of all elements either belonging to U or lying above (that is, $\geqq$) some element of U; dually, define the subset P^-. Both sets P^+ and P^- are non-empty, since both contain U. From the assumption that U does not contain all maximal elements we infer that $P^- \neq P$, and likewise that $P^+ \neq P$. Furthermore, $P^+ \cap P^- = U$, since no element can lie both above and below U unless it is already in U. Finally, $P^+ \cup P^- = P$, for any element p not in $P^+ \cup P^-$ would be unrelated to each of the elements of U, and hence the set $p \cup U$ would be an unrelated subset. This would contradict the assumption that U is a maximal unrelated subset.

Now apply the induction hypothesis to both ordered sets P^+ and P^-. In each of these sets the maximum size of an unrelated set is again n, and therefore there are partitions of P^+ into blocks $B_1, B_2, \ldots, B_n$ and of P^- into blocks $C_1, C_2, \ldots, C_n$, where all the blocks are linearly ordered subsets. Every such block contains a unique element of U; using this element, we pair each block B_i with a unique block, say C_i – after reindexing the blocks C_i if necessary – and the chains $B_1 \cup C_1, B_2 \cup C_2, \ldots, B_n \cup C_n$ are a partition of P into n blocks, as desired.

Case 2. If Case 1 fails, then every unrelated subset of size n either contains all maximal elements of P – if so, it must coincide with the set of all maximal elements of P – or else it similarly coincides with the set of all minimal elements (note that an element of P can be both minimal and maximal). Let a be a maximal element, and let b be a minimal element such that $b \leqq a$. These elements need not be distinct. Let C_1 be the set containing both a and b, and no other elements of P. If n is the maximum size of an unrelated set of P, then the similar number for Q is $n-1$: for if this number were n, we could find an unrelated subset of size n of P containing some element which is neither maximal nor minimal, which would send us back to Case 1. By the inductive hypothesis, there is a partition of Q into $n-1$ chains, call them $C_2, C_3, \ldots, C_n$. The chains $C_1, C_2, \ldots, C_n$ are a partition of P, q.e.d.

4. The Marriage Theorem

With every relation R of sets A and B, that is, $R \subseteq A \times B$, we can associate an oriented bipartite graph G, whose vertices are the elements of $A \cup B$, and whose oriented edges are in one-to-one correspondence with the ordered pairs $(a, b) \in R$. Such a graph G is the Hasse diagram of an ordered set $P = A \cup B$, so defined that A is the set of maximal elements, and B is the set of minimal elements. We shall blithely identify R, G and P, and talk interchangeably about 'edges of a relation', 'chains in R', 'the ordered set $A \cup B$', etc. The ensuing confusion turns out to be useful and instructive.

When can one find a subset, or sub-relation $F \subseteq R$, such that F is the graph of a one-to-one function from A to B? Say that the relation R has a *matching* when such a subrelation F can be found. The best-known answer is provided by the

Marriage Theorem. A necessary and sufficient condition that a relation $R \subseteq A \times B$ between finite sets A and B have a matching is that, for every positive integer k, every k-subset of A be related to a subset of B having at least k elements.

A subset K of A is said to be *related* to a subset L of B whenever L contains all the heads of oriented edges whose tails are vertices in A.

As in Dilworth's Theorem, the condition is obviously necessary. To prove it sufficient, we first show that B is an unrelated subset of maximum size in the ordering defined by the relation. Once this is established, we apply Dilworth's Theorem and conclude that since every element of B must belong to a chain in any minimal partition of $A \cup B$, every element of A belongs to a chain containing exactly two elements. These chains supply the desired matching.

Now let us prove that B is a maximal unrelated set. Supposing this were not so, let U be a maximal unrelated set of larger size than B, say n, so that $\nu(B)-n < 0$. Because $U = (U \cap A) \cup (U \cap B)$, we have $\nu(U \cap B) = n - \nu(U \cap A)$, and

$$\nu(B - U \cap B) = \nu(B) - \nu(U \cap B) = \nu(B) - [n - \nu(U \cap A)]$$

whence

$$\nu(B - U \cap B) = \nu(B) - n + \nu(U \cap A) < \nu(U \cap A),$$

since $\nu(B) - n < 0$. But the set $U \cap A$ can only be related to a subset of $B - U \cap B$, otherwise U would contain two related elements. The above inequality states that the set $U \cap A$ is related to a set having fewer than $v(U \cap A)$ elements, and this contradicts the hypothesis of the Theorem, q.e.d.

The König-Egervary Theorem. In a bipartite graph, a *cut-set* is a set of vertices which meets every edge. The following variant of the Marriage Theorem is often useful:

Theorem 4.2. The minimal size of a cut-set of a finite bipartite graph equals the maximal size of a family of disjoint chains, that is, the maximal size of a set of edges no two of which meet.

The proof results from two steps:

(a) If C is a minimal cut-set, then the complement of C is a maximal set of unrelated elements in the partial ordering. Indeed, every edge of the graph has at least one vertex in the cut-set C, and thus no two vertices in the complement of C can be the end points of the same edge.

(b) Let E be a maximal-size family of edges, no two of which meet. The edges of E, plus enough blocks with one element to include all vertices not belonging to aned ge in E, give a minimal partition π of the ordered set into chains. But the number of edges in the maximal family E, plus the number of blocks of the partition π, equals n, the number of vertices of the bipartite graph.

Thus, as in step (a), the number of blocks in a minimal partition into chains equals n minus the number of edges in E. The proof is concluded upon comparing steps (a) and (b) and applying Dilworth's Theorem.

The König-Egervary Theorem can be restated in terms of matrices of zeros and ones. Call a *line* of a matrix either a row or a column (a felicitous term due to H. Ryser). Then:

Theorem 4.3. The minimum size of a family of lines, in a matrix of zeros and ones, subject to the condition that every non-zero entry of the matrix be contained in a line of the family, equals the maximum size of a set of entries of the matrix, no two of which lie on the same line.

With a matrix (x_{ab}) of zeros and ones associate a relation R between the set A of rows and the set B of columns, by relating a row a to a column b whenever the entry x_{ab} of the matrix is one. The above statement then becomes a rephrasing of the König-Egervary Theorem.

One can reverse the process, and starting with a relation R between A and B, obtain a matrix of zeros and ones in the manner indicated. In this way, every statement of matching theory can be translated into a theorem about matrices of zeros and ones. We shall see on another occasion that there is another way of associating a matrix with a relation, which in some respects is more fruitful.

Example. Given an n by n chessboard, amputated as indicated in Section 1. When is it possible to cover it by 'dimers'? An answer is provided by the Marriage Theorem. After labeling adjacent squares either white or black, one obtains a relation R between the set W of white squares and the set B of black squares. Two squares are related whenever they are adjacent (that is, whenever they have a boundary edge in common). The Marriage Theorem gives a necessary and sufficient condition that the board be coverable by dimers.

In practice, the verification of the conditions of the Marriage Theorem in this case can be considerably simplified. Roughly speaking, one need not consider all subsets of W – or of B – but only those near the boundary of the board.

5. A Coding Theorem

A family F of subsets of a finite set S is to be faithfully represented, as an ordered set, by a set of n-vectors $(x_1, x_2, \ldots, x_n)$ with integer entries, with the componentwise ordering. How does the minimal dimension n of such a representation depend upon the structure of the family F? Call such a representation a *coding* of the family F; our problem is to find the most economical coding. The minimum dimension n of a coding is a useful measure of the complexity of the family F, closely related to the notion of *entropy*.

A familiar example of a coding is the representation of the family of all subsets of an n-set S by the set of all n-vectors whose entries are zeros and ones. It is not immediately clear whether one can do better by allowing the entries to range over the integers. The answer is trivially obtained as an application of the results we proceed to derive. Any coding of a family F of subsets extends to a coding of the larger family $L(F)$ obtained by closing F under union and intersection – but not complementation! – Every such family of sets $L(F)$ forms a distributive lattice, and it is in fact convenient to start with an abstract distributive lattice, namely, a lattice whose join and meet satisfy the distributive laws

$$a \wedge (b \vee c) = (a \wedge b) \vee (a \wedge c)$$

$$a \vee (b \wedge c) = (a \vee b) \wedge (a \vee c).$$

Incidentally: it is an amusing exercise in elementary combinatorics to show that either of the two above laws implies the other.

The reader should also convince himself that the analogous distributive laws with several 'factors' follow from either of the above two.

A sublattice of a product of distributive lattices is again a distributive lattice; thus, any ordered set of vectors with integer components, which is closed under the operations of taking componentwise maxima and minima, is a distributive lattice.

We define a *coding* of a distributive lattice L as an isomorphism of L into a product $C_1 \times C_2 \times \ldots \times C_n$, preserving joins and meets. Each of the C_i is a set $[0, 1, \ldots, c_i]$ of integers in their natural order.

The economical codings of distributive lattices were characterized by Dilworth in the beautiful

Theorem 5.1. The minimal dimension of a coding of a finite distributive lattice L equals the maximal number of elements covering any single element of L.

The proof will lead us through a number of lemmas and a detour into the combinatorics of distributive lattices, much of which is of independent interest.

An element q of a lattice L is said to be *join-irreducible* if $q = x \vee y$ implies either $q = x$ or $q = y$. Clearly, every element of a finite lattice is the join of a set of join-irreducibles; our first aim is to show that for a finite distributive lattice this set is unique. A meet-irreducible (element) is dually defined.

Lemma 1. If q is a join-irreducible element of a distributive lattice, then $q \leqq x \vee y$ implies $q \leqq x$ or $q \leqq y$.

Proof. By the distributive law, $q = q \wedge (x \vee y) = (q \wedge x) \vee (q \wedge y)$. By definition of a join-irreducible, we must have either $q = q \wedge x$ or $q = q \wedge y$, and this is precisely the conclusion of the lemma.

Lemma 1 generalizes to $q \leqq x \vee y \vee \ldots \vee z$, with an arbitrary finite number of elements on the right side of the inequality. When referring to Lemma 1, we assume that the statement has been thus extended.

In a lattice L, every element which is the join of a finite subset of join-irreducibles, is also the join of a finite *unrelated* set of join-irreducibles (simply discard the unneeded elements). If L is distributive, we can say more:

Lemma 2. Let P be the set of join-irreducibles of a finite distributive lattice L. Then every element of L is the join of a unique unrelated subset of P.

Proof. Only the uniqueness is to be proved. Suppose

$$x = q_1 \vee q_2 \vee \ldots \vee q_k = p_1 \vee p_2 \vee \ldots \vee p_n, \qquad n \geqq k,$$

where both the q's and the p's are unrelated sets of join-irreducibles. By the distributive law

$$q_i = q_i \wedge x = (q_i \wedge p_1) \vee (q_i \wedge p_2) \vee \ldots \vee (q_i \wedge p_n).$$

By Lemma 1, we have $q_i = q_i \wedge p_j$ for some index j. In other words, for every q_i we can find a p_j such that $q_i \leqq p_j$. The same argument shows that for every p_j we can find a q_m such that $p_j \leqq q_m$. The two inequalities combined give

$$q_i \leqq p_j \leqq q_m.$$

Any two of the q's are unrelated, and this can happen only if $i = m$; hence, $q_i = q_m = p_j$, q.e.d.

An *ideal* I in an ordered set P is a subset of P such that, if $x \in I$ and $y \leqq x$, then $y \in I$. The set-theoretic union or intersection of two or more ideals is again an ideal. Therefore, the family of all ideals of an ordered set P is a distributive lattice, which we denote by $I(P)$. It turns out that every distributive lattice is isomorphic to an $I(P)$. This result, due to Birkhoff, is far more important than the classical theorem of Boole, popularized in the current frenzy for the 'new math', stating that every finite Boolean algebra is isomorphic to the lattice of all subsets of a finite set. Much more is true, namely, every finite distributive lattice is isomorphic to a family of subsets of a finite set:

Theorem 5.2 (Birkhoff). Let P be the ordered set of join-irreducibles of a finite distributive lattice L. Then L is isomorphic to the distributive lattice $I(P)$ of all ideals of P.

Proof. By Lemma 2, to every element x of L there corresponds a unique minimal set $S(x)$ of join-irreducibles, and conversely. This gives a one-to-one correspondence between L and the family of all unrelated subsets of P. On the other hand, to every unrelated subset $S(x)$ of P there corresponds the unique ideal $I(x)$ of P, 'generated by $S(x)$', that is, $I(x) = \{p \in P: p \leqq q \text{ for some } q \in S(x)\}$. Conversely, to every ideal I of P there corresponds a set $S(x)$ of unrelated elements, namely, the set of maximal elements of I. Therefore, every ideal I corresponds biuniquely to an element x of L, or $I = I(x)$. To complete the proof, we must show that the map $x \to I(x)$ between the distributive lattice L, and the distributive lattice of ideals of P, preserves order.

Suppose that $x \leqq y$ in L, and let $p \in I(x)$. By the distributive law,

$$p = p \wedge y = p \wedge (\vee S(y)) = \bigvee_{u \in S(y)} (p \wedge u).$$

Since p is join-irreducible in L, it follows from Lemma 1 that $p \leqq u$ for some u in $S(y)$, that is, that $p \in I(y)$. Thus $I(x) \subseteq I(y)$.

Conversely, suppose that $I(x) \subseteq I(y)$. To every $u \in S(x)$ there is a $v \in S(y)$ such that $u \leqq v$; hence $\vee S(x) \leqq \vee S(y)$, or $x \leqq y$, q.e.d.

Corollary 1. Two distributive lattices are isomorphic if and only if their ordered sets of join-irreducibles are isomorphic.

The proof is immediate.

The word 'isomorphism' in Theorem 5.2 has been purposely left imprecise. The Theorem remains correct, whether the word is interpreted as 'lattice isomorphism', or as 'order isomorphism'. For distributive lattices, either kind of isomorphism implies the other.

Theorem 5.2 shows that there are as many non-isomorphic distributive lattices as there are non-isomorphic ordered sets. We shall briefly detour to describe another application of the Theorem.

Corollary 2. Let H be the Hasse diagram of a finite distributive lattice. For any positive integer n, the number of vertices of H which are tails of exactly n edges equals the number of vertices of H which are heads of exactly n edges.

For example, for $n = 1$ the Corollary states that the number of join-irreducibles of a distributive lattice equals the number of meet-irreducibles. We require the preliminary

Lemma 3. An element of a finite distributive lattice covers exactly n elements whenever it is the join of an unrelated set of n join-irreducibles.

Proof. Represent the lattice as $I(P)$, as in Theorem 1. If I is an ideal of P, and p a maximal element of I, then the set $I-p$ is an ideal, and I covers $I-p$ in the lattice $I(P)$. If J is an ideal and $J \subsetneq I$, then J omits at least one maximal element p of I, so that $J \subseteq I-p$. If I covers J, then $J = I-p$. Therefore, I covers as many ideals in $I(P)$ as it has maximal elements. But I is the union of as many unrelated join-irreducibles as it has maximal elements, by Theorem 1; this concludes the proof.

Now to the proof of Corollary 2. Again we represent L as $I(P)$, and the dual lattice L^* as $I(Q)$, where Q is the set of meet-irreducibles of L. The map $I \to I'$, where I' is the set-theoretic complement of the ideal I, gives an order-antiisomorphism of $I(P)$ with $I(P^*)$, so that $I(P^*)$ is isomorphic with $I(Q)$. Hence, by Corollary 1, the set P of join-irreducibles of $I(P)$ is antiisomorphic to the set Q of meet-irreducibles. Thus, there are as many unrelated sets of n join-irreducibles, as there are unrelated sets of n meet irreducibles. The conclusion now follows from the preceding Lemma.

We now have all the tools to conclude the proof of the Coding Theorem stated at the beginning.

Proof of Theorem 5.1. Again we identify the distributive lattice L with the lattice $I(P)$ of ideals of the ordered set P of join-irreducibles of L. Every element x of L is thus a *subset* of P, and meets and joins in L are to be identified with unions and intersections of subsets of P.

Let k be the maximum size of a set of elements covering a single element x of L, and let n be the minimum dimension of a coding. Lemma 3 shows that $n \geqq k$, and we prove the converse inequality.

By Dilworth's Theorem, the ordered set P can be partitioned into k disjoint chains $C_1, C_2, \ldots, C_k$. Let $p_i = x \cap C_i$; note that p_i may be the null set. Because the chains C_i cover the set P, we have that $x = p_1 \cup p_2 \cup \ldots \cup p_k$; if $q_i = y \cap C_i$ for some other element $y \in L$, then $(x \cup y) \cap C_i = (x \cap C_i) \cup (y \cap C_i) = p_i \cup q_i$ by the distributive law, and $(x \cap y) \cap C_i = (x \cap C_i) \cap (y \cap C_i) = p_i \cap q_i$. Furthermore, the sets p_i are chains, therefore, if $p_i = q_i$ for all i, then $x = y$. We have thus shown that the mapping

$$x \to (p_1, p_2, \ldots, p_k)$$

is an isomorphism of the lattice L into the product $C_1 \times C_2 \times \ldots \times C_k$.

To complete the proof, replace C_i by the chain N_i of integers $0, 1, 2, \ldots, v(C_i)$, and map p_i into the integer $v(p_i)$. This gives an isomorphism of $C_1 \times C_2 \times \ldots \times C_k$ onto the ordered set $N_1 \times N_2 \times \ldots \times N_k$, q.e.d.

6. Doubly Stochastic Matrices

We prove a beautiful result discovered by Garrett Birkhoff, and first published in an obscure Argentine periodical. John von Neumann rediscovered it independently some time later, but never published his proof, which was transmitted by word-of-mouth, and which is quite different from the proof given below. The topic lies at the borderline of matrix algebra and convexity.

The Birkhoff-von Neumann Theorem has motivated a small flood of work in linear inequalities and in functional analysis.

A *doubly stochastic* matrix $X = (x_{ab})$ is a matrix with non-negative entries, with the property that the sum of the entries on each line (row or column) is equal to one. A doubly stochastic matrix is evidently a square matrix. The set D of all n by n doubly stochastic convex, when considered as a set of points in real affine space of matrices is dimension n^2. Every permutation matrix (namely, a doubly stochastic matrix all of whose entries are either 0 or 1) is an extremal point of the convex set D. Recall that an *extremal point* of a convex set is one that is not a convex combination of any finite subset; a point x in real affine space is a *convex combination* of points $u, v, \ldots, w$, when

$$x = \lambda u + \mu v + \ldots + \nu w,$$

where $\lambda, \mu, \ldots, \nu$ are non-negative reals, none of them equal to one, and $\lambda + \mu + \ldots + \nu = 1$.

Are the permutation matrices the only extremal points? The answer is obtained by an application of the Marriage Theorem:

Theorem 6.1. (*Birkhoff-von Neumann*). Every doubly stochastic matrix is a convex combination of permutation matrices.

Proof. Associate with a doubly stochastic matrix $X = (x_{ab})$ a relation R of the set A of rows of X, and the set B of columns of X, relating an element a of A to an element b of B, whenever the entry x_{ab} of the matrix is non-zero. The relation R thus defined satisfies the condition of the Marriage Theorem. Suppose not. Then a subset J of rows is related to a set K of columns of smaller size. We visualize the situation in the picture below, where the submatrix with rows J and columns K is shaded:

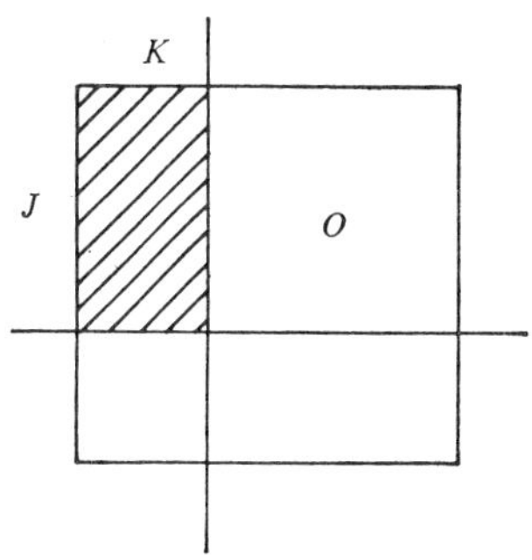

Fig. 4.1.

If $\nu(J) > \nu(K)$, then at least one column in the set K must have entries which add up to more than one, contradicting the definition of a doubly stochastic matrix.

Now use induction on the number of non-zero entries of X. If X has n non-zero entries – it cannot have fewer –, then X is already a permutation matrix. If X has more than n non-zero entries, apply the Marriage Theorem to the relation R, and conclude that there is a permutation matrix $P = (p_{ab})$, with the property that $p_{ab} = 1$ only if $x_{ab} > 0$; let ε be the value of the smallest of the entries x_{ab} of X for which $p_{ab} = 1$. Then the matrix $Y = X - \varepsilon P$ has non-negative entries, and the sum of the entries on each line of Y is $1-\varepsilon$; the matrix $(1-\varepsilon)^{-1}Y$ is doubly stochastic. Furthermore, the matrix X is a convex combination of the doubly stochastic matrices P and $(1-\varepsilon)^{-1}Y$, because

$$X = \varepsilon P + (1-\varepsilon)(1-\varepsilon)^{-1}Y.$$

If the matrix $(1-\varepsilon)^{-1}Y$ is a permutation matrix, the proof is concluded; otherwise, the matrix $(1-\varepsilon)^{-1}Y$ is a doubly stochastic matrix having at least one less non-zero entry than X, namely, any entry of X which takes the value ε. This completes the induction step and concludes the proof. The following result, due to König, is an immediate consequence of the Birkhoff-von Neumann Theorem:

Corollary. Let X be a matrix of zeros and ones, with the property that the sum of the entries on each line is constant. Then X is the sum of permutation matrices.

7. Inequalities for Symmetric Functions

The best-known of all classical inequalities is the inequality of the arithmetic and geometric means:

$$\sqrt[n]{x_1 x_2 \ldots x_n} \leqq (x_1 + x_2 + \ldots + x_n)/n,$$

where the x_i are non-negative reals, and equality holds whenever all the x_i are equal. In 1903, R. F. Muirhead published a far-reaching generalization of this inequality. We derive Muirhead's result from the Birkhoff-von Neumann Theorem.

A vector $\boldsymbol{a} = (a_1, a_2, \ldots, a_n)$ with non-negative components is said to be an *average* of the vector $\boldsymbol{b} = (b_1, b_2, \ldots, b_n)$, also with non-negative components, when there is a doubly stochastic matrix X such that $\boldsymbol{a} = X\boldsymbol{b}$. For example, the vector $(1/n, 1/n, \ldots, 1/n)$ is an average of the vector $(1, 0, 0, \ldots, 0)$, obtained by applying the doubly stochastic matrix all of whose entries equal $1/n$. We use the notation

$$[\boldsymbol{a}] = [a_1, a_2, \ldots, a_n] = \frac{1}{n!}\sum_{\sigma} x_1^{a_{\sigma 1}} x_2^{a_{\sigma 2}} \ldots x_n^{a_{\sigma n}}$$

where the sum on the right ranges over all permutations σ of the subscripts $1, 2, \ldots, n$, and call such an expression a *symmetrical mean*. For example, the symmetrical means corresponding to the vectors $\boldsymbol{a}$ and $\boldsymbol{b}$ of the preceding example are the arithmetic and geometric means of the x_i. The symmetrical mean $[1, 2, 0, \ldots, 0]$ is the polynomial

$$\frac{1}{n(n-1)} \sum_{i,j} x_i x_j^2 .$$

Theorem 7.1. (*Muirhead's inequality*). Given two symmetrical means $[\boldsymbol{a}]$ and $[\boldsymbol{b}]$, the inequality

$$[\boldsymbol{a}] \leqq [\boldsymbol{b}]$$

holds for all non-negative values of the variables x_i, if and only if the vector $\boldsymbol{a}$ is an average of the vector $\boldsymbol{b}$.

Proof. The condition is sufficient. Indeed, let $\boldsymbol{a} = X\boldsymbol{b}$, where the matrix X is doubly stochastic, and express the matrix X as a convex combination of doubly stochastic matrices:

$$X = \sum_P \lambda_P P$$

where P is a permutation matrix, the λ_P are non-negative reals adding up to 1, and the sum ranges over all permutation matrices P.

Let $\boldsymbol{y} = (\log x_1, \log x_2, \ldots, \log x_n)$. Assume that none of the x_i vanishes – the general case follows immediately by continuity. Write $(\boldsymbol{c}, \boldsymbol{y}) = \sum_i c_i y_i$. We have then

$$[\boldsymbol{a}] = (\sum_Q \exp(Q\boldsymbol{a}, \boldsymbol{y}))/n!$$

where again the variable Q ranges over the set of permutation matrices. Since $\boldsymbol{a} = X\boldsymbol{b}$, we can write

$$(Q\boldsymbol{a}, \boldsymbol{y}) = \sum_P \lambda_P (QP\boldsymbol{b}, \boldsymbol{y}).$$

We now use the fact that the function e^x of the real variable x is convex (this is the main step in the proof), namely, that

$$\exp \sum_i \lambda_i x_i \leqq \sum_i \lambda_i e^{x_i}; \qquad \lambda_i \geqq 0, \quad x_i \geqq 0, \quad \sum_i \lambda_i = 1.$$

This is verified by a simple – and, hopefully, well-known – argument on the graph of the function e^x. We obtain

$$\exp(\sum_P \lambda_P (QP\boldsymbol{b}, \boldsymbol{y})) \leqq \sum_P \lambda_P \exp(QP\boldsymbol{b}, \boldsymbol{y}) \qquad (*)$$

whence

$$n![\boldsymbol{a}] \leqq \sum_Q \sum_P \lambda_P \exp(QP\boldsymbol{b}, \boldsymbol{y}) = \sum_P \lambda_P \sum_Q \exp(QP\boldsymbol{b}, \boldsymbol{y}).$$

But

$$\sum_Q \exp(QP\boldsymbol{b}, \boldsymbol{y}) = \sum_Q \exp(Q\boldsymbol{b}, \boldsymbol{y}) = [\boldsymbol{b}]n!$$

hence

$$[\boldsymbol{a}] \leqq \sum_P \lambda_P[\boldsymbol{b}] = [\boldsymbol{b}],$$

since $\sum_P \lambda_P = 1$. This completes the proof that the condition is sufficient.

The condition is necessary. Indeed, let $x_1 = x_2 = \ldots = x_k = x$, and set all other x_i equal to 1. Then the means $[\boldsymbol{a}]$ and $[\boldsymbol{b}]$ become polynomials in the variable x. The inequality will hold for large x only if the power of x occurring on the left side does not exceed the power of x occurring on the right side. If we assume – as we may – that

$$a_1 \geqq a_2 \geqq \ldots \geqq a_n \geqq 0; \qquad b_1 \geqq b_2 \geqq \ldots \geqq b_n \geqq 0 \qquad (*)$$

then this remark gives the conditions

$$a_1+a_2+\ldots+a_k \leqq b_1+b_2+\ldots+b_k, \qquad 1 \leqq k \leqq n. \qquad (**)$$

Furthermore, letting all x_i equal x, and remarking that the resulting inequality between two polynomials in x must also hold for small positive values of x, we obtain the further condition

$$a_1+a_2+\ldots+a_n = b_1+b_2+\ldots+b_n. \qquad (***)$$

The rest of the proof of necessity is a consequence of the

Lemma. Let the vectors $\boldsymbol{a}$ and $\boldsymbol{b}$ satisfy conditions (*), (**) and (***). Then there exists a doubly stochastic matrix X such that $\boldsymbol{a} = X\boldsymbol{b}$.

Proof. Since every doubly stochastic matrix is a convex combination of permutation matrices, $\boldsymbol{a} = X\boldsymbol{b}$ if and only if the vector $\boldsymbol{a}$ lies in the convex hull of the set of vectors obtained from $\boldsymbol{b}$ by permuting the components of $\boldsymbol{b}$ in all possible ways. We prove this latter statement by induction on the dimension n.

If the vectors $\boldsymbol{a}$ and $\boldsymbol{b}$ have one equal component, say $a_i = b_i$, then the $(n-1)$-dimensional vectors obtained by removing the i-th components again satisfy the three conditions, and the conclusion follows from the induction hypothesis. We may and shall therefore assume that no two components of $\boldsymbol{a}$ and $\boldsymbol{b}$ coincide. We show that a vector $\boldsymbol{c}$ can be found, which majorizes $\boldsymbol{a}$ in the sense of the Lemma, is a convex combination of the permutations of $\boldsymbol{b}$, and has one component in common with $\boldsymbol{a}$. Since a convex combination of permutations of $\boldsymbol{c}$ is also a convex combination of permutations of $\boldsymbol{b}$, the proof will be complete.

Because $a_1 < b_1$ and $\sum_i(a_i - b_i) = 0$, there is an index i for which $a_i < b_i$ and $a_{i+1} > b_{i+1}$.

Case 1. $a_i \geqq (b_i + b_{i+1})/2$. Set $c_i = a_i$, $c_{i+1} = b_{i+1} + b_i - c_i$, and all other $c_j = b_j$. The vector $\boldsymbol{c}$ satisfies the three conditions of the Lemma and is a convex combination of $\boldsymbol{b}$ and the vector obtained from $\boldsymbol{b}$ by permuting the components b_i and b_{i+1}.

Case 2. If Case 1 fails, then $a_{i+1} < a_i < (b_i + b_{i+1})/2$. Set $c_{i+1} = a_{i+1}$, $c_i = b_i + b_{i+1} - c_{i+1}$, and all other $c_j = b_j$. Again, the vector $\boldsymbol{c}$ satisfies the three conditions of the Lemma, and is a convex combination of $\boldsymbol{b}$ and the vector obtained by permuting b_i and b_{i+1}, q.e.d.

It is left to the reader to show that the converse of the Lemma also holds. Thus, conditions (*), (**) and (***) give another necessary and sufficient condition for Muirhead's inequality.

As a typical application, we derive the

Corollary. Let $[\boldsymbol{a}]$ be an average of the variables $x_1, x_2, \ldots, x_n$, and let the sum of the components of the vector $\boldsymbol{a}$ be one. Let A be the arithmetic mean, and let G be the geometric mean. Then

$$G \leqq [\boldsymbol{a}] \leqq A.$$

Proof. The vector $\boldsymbol{g} = (1/n, 1/n, \ldots, 1/n)$, whose average $\boldsymbol{g} = G$ is the geometric mean, equals the convex combination

$$\boldsymbol{g} = \frac{1}{n!} \sum_P P\boldsymbol{a} \tag{\#}$$

where the sum again ranges over all permutation matrices P. For the right side is invariant under any permutation, and the vector $\boldsymbol{g}$ is the only one in the convex set which is invariant under all permutations. The left inequality of the Corollary follows from (#) upon applying Muirhead's inequality.

The right inequality is even easier. If $\boldsymbol{b} = (1, 0, 0, \ldots, 0)$, so that $[\boldsymbol{b}] = A$, then the vectors $P\boldsymbol{b}$, as P ranges over the set of all permutation matrices, are extremals of the convex set consisting of all vectors with non-negative coordinates whose sum is one. Thus $\boldsymbol{a}$ is a convex combination of the $P\boldsymbol{b}$ and the conclusion again follows from Muirhead's inequality.

Equality holds in Muirhead's inequality whenever it holds in inequality (*), used in the course of the proof, and this happens only if all components of the vector $\boldsymbol{b}$ are identical. This is a consequence of the fact that the function e^x is *strictly* convex.

8. Doubly Stochastic Measures

The problem of obtaining continuous analogs of the results of matching theory has been often stated, but little worked upon. Here, as elsewhere, the difficulty lies in singling out the significant problems, and in achieving definitions which are general enough without running into triviality. Two alternatives arise in deriving continuous structures by analogy with the discrete: one can use the ideas of measure theory, or else one can introduce topological notions. A successful instance of the first alternative is the solution of Birkhoff's celebrated Problem 111, recently obtained almost simultaneously and independently by R. Douglas and J. Lindenstrauss, by a short argument drawn from functional analysis. Their result paves the way to the development of a 'measure-theoretic' matching theory.

Birkhoff's Problem 111 asked, somewhat loosely, for a 'continuous analog' of the Birkhoff-von Neumann Theorem. Now, the continuous analog of a doubly stochastic matrix is a doubly stochastic measure (v. below). The difficulty lay in guessing the correct characterization of all extremal doubly stochastic measures, in view of the non-existence of a measure-theoretic counterpart to the notion of a permutation matrix. It turns out that extremal doubly stochastic measures can be characterized by an approximation property of the spaces of integrable functions they define.

Recall that the notation $\boldsymbol{S} = (S, \mu)$ for a measure space $\boldsymbol{S}$ denotes a set S, and a countably additive positive measure μ defined on a σ-field of subsets of S, which need not be indicated. We assume that every measure space appearing below is finite.

Let F be a linear subspace of real functions contained in $L_1(S, \mu)$, and assume that constant functions belong to F. The set $C(\mu)$ of all positive measures ν with the property that

$$\int f \, d\nu = \int f \, d\mu$$

for all functions f in the subspace F, is convex, as is trivial to verify. The extremals of this set are given by

Theorem 8.1. A measure ν is extremal in the convex set $C(\mu)$ if and only if the subspace F is dense in $L_1(S, \nu)$.

Proof. If the measure ν is not extremal, then (recall the definition of an extremal point of a convex set given in Section 5) $\nu = (\alpha+\beta)/2$, where α and β are two distinct measures in the set $C(\mu)$. Therefore, each of the measures α and β is absolutely continuous relative to the measure ν. By the Radon-Nikodym Theorem, we can write α in the form

$$\alpha = h \, d\nu$$

where h is a non-negative bounded function. Since α is assumed not to coincide with ν, the function h is not identically equal to 1, that is, $1-h \neq 0$. If so, then for every function f in the subspace F we have

$$\int f(1-h) d\nu = \int f \, d\nu - \int f h \, d\nu = \int f \, d\nu - \int f \, d\alpha = 0,$$

because $\alpha \in C(\mu)$. In other words, the function $1-h$ is orthogonal to the subspace F, and so by the Hahn-Banach Theorem the subspace F is not dense in $L_1(S, \nu)$.

Conversely, assume that the subspace F is not dense in $L_1(S, \nu)$. We prove that the measure ν is not extremal, that is, we find two distinct measures in the convex set $C(\mu)$, such that $\nu = (\alpha+\beta)/2$. By the Hahn-Banach Theorem, we can find a bounded function h which is not zero and vanishes on F:

$$\int f h \, d\nu = 0, \qquad f \in F$$

We may and shall assume that $||h||_\infty = 1$. Define a measure λ by the indefinite integral

$$\lambda(E) = \int_E h \, d\nu, \qquad E \in \Sigma$$

and set

$$\alpha = \nu + \lambda, \qquad \beta = \nu - \lambda.$$

Since $h \leqq 1$, both α and β are positive. The function h is not identically equal to one, because the subspace F is assumed to contain constant functions; therefore, $\lambda \neq \nu$. From the above equations it follows that $\nu = (\alpha+\beta)/2$, so all we have to verify is that the measures α and β belong to the convex set $C(\mu)$. This follows from the following string of identities, valid for all functions $f \in F$:

$$\int f \,\mathrm{d}\alpha = \int f \,\mathrm{d}\nu + \int f \,\mathrm{d}\lambda = \int f \,\mathrm{d}\nu + \int f h \,\mathrm{d}\nu = \int f \,\mathrm{d}\nu = \int f \,\mathrm{d}\mu,$$

q.e.d.

The preceding Theorem has several interesting applications. Define a doubly stochastic measure ν on the product $(S, \sigma) \times (T, \nu)$ two probability spaces (S, σ) and (T, τ) with measures σ and τ defined on sets S and T, by the property that $\nu(A \times T) = \sigma(A)$ for a measurable set $A \subseteq S$, and $\nu(S \times B) = \tau(B)$ for a measurable set $B \subseteq T$. The solution of Birkhoff's Problem 111 is given by:

Theorem 8.2. Let D be the set of all doubly stochastic measures on the product $(S \times T, \sigma \times \tau)$ of two probability spaces (S, σ) and (T, τ). A measure ν is extremal in the set D whenever real functions of the form $f(s) + g(t)$, where $s \in S$ and $t \in T$, are dense in the product space $L_1(S \times T, \nu)$.

Proof. Let F be the subspace of $L_1(S \times T, \sigma \times \tau)$ consisting of all functions of the form $f(s) + g(t)$. The conclusion of the Theorem will follow at once from Theorem 1, if we show that a measure ν is doubly stochastic if and only if

$$\int (f+g)\mathrm{d}\nu = \int (f+g)\mathrm{d}(\sigma \times \tau)$$

for all $f + g \in F$. This in turn follows from an approximation argument: the identity holds if f and g are indicators (= characteristic functions) of measurable sets – this is just the definition of a doubly stochastic measure –; hence, it also holds if f and g are simple functions; finally, it holds if f and g are limits of simple functions, q.e.d.

As an example, let S and $\varDelta$ be atomic probability spaces, each having n atoms of measure $1/n$. The verification that an extremal doubly stochastic measure in the above sense is a permutation matrix amounts to the von Neumann proof of the Birkhoff-von Neumann Theorem. We leave it to the reader.

A generalization of the same argument can be used to show that all extremal doubly stochastic measures in a non-atomic probability space are singular.

9. Sperner's Theorem and its Extensions

More than 40 years ago, E. Sperner proved that the maximum size of a family of subsets of a set of n elements, subject to the requirement that no set in the family contain another, is the same as the maximum of the binomial coefficients $\binom{n}{k}$ as $0 \leqq k \leqq n$. This maximum is attained when $k = [n/2]$. We prove a more general theorem. An advantage of the present approach is that it yields similar – and deeper – results when the lattice of subsets is replaced by the lattice of subspaces of a finite-dimensional vector space over a Galois field.

We begin by deriving an explicit partition of the lattice $L(S)$ of subsets of an n-set S into

$$\binom{n}{\left[\frac{n}{2}\right]}$$

chains. This partition can also be used to derive Theorem 1 below, but we shall rederive this Theorem by a more general method.

Theorem 9.1. The lattice of subsets of an n-set can be so partitioned into chains that the number of blocks with k elements equals

$$\binom{n}{[(n+k)/2]} - \binom{n}{[(n+k+1)/2]}$$

Proof. Proceeding by induction, after remarking that the statement is trivial for $n = 1$, assume that it is true for the subset $S-a$, where a is an arbitrary element of S, that is, that the lattice $L(S-a)$ has been partitioned into

$$j = \binom{n-1}{[(n-1)/2]}$$

chains, say $\tilde{C}_1, \tilde{C}_2, \ldots, \tilde{C}_j$. Let c_i be the maximal element of the chain

$\tilde{C}_i$, and define chains (some of which may be empty)

$$C_i = \tilde{C}_i \cup \{c_i \cup \{a\}\},^* \quad 1 \leqq i \leqq j$$

$$C_{i+j} = \{c \cup \{a\} : c \in \tilde{C}_i - c_i\},^{**} \quad 1 \leqq i \leqq j.$$

It is easily verified that the non-void sets among the chains C_i are distinct and are the $\binom{n}{[n/2]}$ blocks of a partition of the lattice $L(S)$.

We shall now pass on to a different method.

We define an *r-family* of an ordered set P as a subset F of P in which no chain contains more than r elements.

Let F be an r-family in the lattice $L(S)$ of subsets of an n-set S. Every maximal chain in $L(S)$ meets F in at most r distinct elements. If an element $x \in L(S)$ has size k, there are $k!(n-k)!$ maximal chains of $L(S)$ passing through x. Let f_k be the size of the subfamily of F, consisting of all elements of size k. Then we obtain the inequality

$$\sum_{k=0}^{n} f_k k!(n-k)! \leqq (n!) \cdot r,$$

since $n!$ is the total of maximal chains in $L(S)$. Dividing by $n!$ we find

$$\sum_{k=0}^{n} \frac{f_k}{\binom{n}{k}} \leqq r. \tag{*}$$

To simplify the writing, we relabel the binomial coefficients $c_0, c_1, \ldots, c_n$, so that $c_0 \geqq c_1 \geqq c_2 \geqq \ldots \geqq c_n$. For example, $c_0 = \binom{n}{n/2}$ if n is even. We also relabel the f_k accordingly, and call g_k the relabeled sequence. Inequality (*) can now be rewritten as follows:

$$\sum_{k=0}^{n} \frac{g_k}{c_k} \leqq r. \tag{**}$$

Equality holds for the family F for which $g_0 = c_0, g_{r-1} = c_{r-1}, \ldots, g_{r-1} = c_{r-1}$ and $g_k = 0$ for $k \geqq r$. If n is even and r is odd, or if n is odd and r even, equality can only hold for such a family. We claim that if the integers n and r are of opposite parity, equality in (**) can hold if and only if $g_0 = c_0, g_1 = c_1, \ldots, g_r = c_r$, and all other $g_k = 0$. This follows from the

* Of these, $\binom{n-1}{[(n-1)/2]}$ are non-empty.

** Of these, $\binom{n-1}{[(n-1)/2]-1}$ or $\binom{n-1}{[(n-1)/2]+1}$ are non-empty, according as n is odd or even.

Lemma. If n and r are of opposite parity, if $c_i \geqq x_i \geqq 0$ for $0 \leqq i \leqq n$ and if

$$x_0+x_1+\ldots+x_n \geqq c_0+c_1+\ldots+c_{r-1},$$

then

$$\sum_{k=0}^{n} \frac{x_k}{c_k} \geqq r, \tag{9.1}$$

and equality is attained if and only if $x_0 = c_0, \ldots, x_{r-i} = c_{r-i}$ and $x_k = 0$ for all other k.

Proof. Let $x_0, \ldots, x_n$ be a non-negative sequence that minimizes the left side of (9.1). If the conclusion does not hold, then there is at least one index $i < r$ for which $x_i < c_i$. Let i be the largest such index. Then there must be an index $j \geqq r$ for which $x_j > 0$, otherwise the sum of x's could not exceed the sum of the c's, contrary to assumption.

Now set $y_i = x_i+x_j$ and $y_j = 0$ if $x_i+x_j \leqq c_i$, and set instead $y_i = c_i$ and $y_i = x_j-(c_i-x_i)$ if $x_i+x_j > c_i$. In either case set all other $y_k = x_k$.

In either case we have

$$\frac{y_i}{c_i}+\frac{y_j}{c_j} < \frac{x_i}{c_i}+\frac{x_j}{c_j},$$

because $c_i > c_j$. Hence

$$\sum_{k=0}^{n} \frac{y_k}{c_k} < \sum_{k=0}^{n} \frac{x_k}{c_k};$$

this contradicts the assumption that the sequence x_i minimizes the left side of (1) and completes the proof.

If the integers n and r are of the same parity, then there are at least two families F of maximum size, obtained by taking $g_0 = c_0, \ldots, g_{r-1} = c_{r-1}$ and all other $g_k = 0$, or else $g_0 = c_0, \ldots, g_{r-1} = 0$, $g_r = c_r$ and all other $g_k = 0$. In this case it can also be shown that these are the only extremal families. We omit the additional argument necessary to establish this fact.

In conclusion we have proved

Theorem 9.2. In the lattice of subsets of an n-set, every r-family is of size not larger than

$$\binom{n}{[(n+1)/2]}+\binom{n}{[(n+2)/2]}+\ldots+\binom{n}{[(n+r)/2]}.$$

Similar results hold for the lattice of subspaces $L(V)$ of an n-dimensional vector space over a Galois field with q elements. Recall that the number of subspaces of dimension k is the Gaussian polynomial $\left[{n \atop k}\right]$. In analogy with the proof of Theorem 2 we have the inequality

$$\sum_{k=0}^{n} \frac{f_k}{\left[{n \atop k}\right]} \leqq r$$

holding for all r-families F of subspaces. Reasoning as in the proof of Theorem 2 we find that maximum size of an r-family is the sum of the Gaussian polynomials

$$\left[{n \atop [(n+1)/2]}\right] + \left[{n \atop [(n+2)/2]}\right] + \cdots + \left[{n \atop [(n+r)/2]}\right].$$

and there is only one extremal family if n and r are of different parity.

10. The Littlewood-Offord Problem

Let $a_1, a_2, \ldots, a_n$ be complex numbers of absolute value greater than one. How many among the sums

$$\sum_{j=1}^{n} \varepsilon_j a_j \tag{10.1}$$

where $\varepsilon_j = \pm 1$, lie inside the unit circle? This problem was posed in 1943 by Littlewood and Offord in their study of the zeros of random polynomials. Their partial results were improved by Erdös and others. Over the years, the problem has served as one of the standard brain-teasers among combinatorially inclined mathematicians. The solution we present relies upon Theorems 1 and 2 of Section 9, aided by the

Lemma. Let F be a family of subsets of an n-set S, and let T be a subset of S. Assume that if A and B belong to F, if $A \supseteq B$ and if $A-B$ is contained either in T or in the complement of T, then $A = B$. Then the family F has at most $\binom{n}{[n/2]}$ elements.

Proof. Let π be a 'symmetric partition' into chains of the family of subsets of T, with the property stated in Theorem 1 of Section 9. Let $D_1, D_2, \ldots, D_k$ be subsets of T belonging to the same block of π, and let F_i be the family of subsets E of $S-T$ such that $E \cup D_i \in F$. The second assumption of the Lemma implies that no two elements of the family F_i are comparable, and the first, that the families F_i and F_j are disjoint if $i \neq j$. Applying Theorem 2 of Section 9 to the set $S-T$, we infer that the size of the union $F_1 \cup F_2 \cup \ldots \cup F_k$ is at most

$$\sum_{j=1}^{k} \binom{n-t}{[(n-t+j)/2]}$$

where t is the size of T.

By Theorem 1 of Section 9, there are

$$\binom{t}{[(t+k)/2]} - \binom{t}{[(t+k+1)/2]}$$

chains of length k in the partition π of $L(T)$. Adding over all chains of π, it follows that the size of F is at most

$$\sum_{k=1}^{t+1} \sum_{j=1}^{k} \binom{n-t}{[(n-t+j)/2]} \left[\binom{t}{[(t+k)/2]} - \binom{t}{[(t+k+1)/2]} \right].$$

After commuting the order of summation, this expression equals

$$\begin{aligned} \sum_{r=1}^{t+1} \sum_{k=r}^{t+1} \binom{n-t}{[(n-t+r)/2]} &\left[\binom{t}{[(t+k)/2]} - \binom{t}{[(t+k+1)/2]} \right] \\ &= \sum_{r=1}^{t+1} \binom{n-t}{[(n-t+r)/2]} \binom{t}{[(t+r)/2]} \\ &= \sum_{r=1}^{t+1} \binom{n-t}{[(n-t+r)/2]} \binom{t}{t-[(t+r)/2]}. \end{aligned}$$

The last expression on the right equals $\binom{n}{[n/2]}$. Indeed, let N be the family of subsets of S having $\binom{n}{[n/2]}$ elements, and let N_r be the subfamily of N of all subsets which meet T in an $(r-1)$-subset. There are

$$\binom{n-t}{[(n-t+r)/2]} \binom{t}{t-[(t+r)/2]}$$

elements in the family N_r, and N is the disjoint union of the families N_r. The proof is complete.

The Littlewood-Offord problem is now easily solved:

Theorem 3. (Kleitman) Let $a_1, a_2, \ldots, a_n$ be complex numbers of absolute value greater than one. The number of sums

$$\sum_{j=1}^{n} \varepsilon_j a_j \qquad \varepsilon_j = \pm 1 \tag{1}$$

which lie inside any circle of unit radius, is at most $\binom{n}{[n/2]}$.

Proof. Since the statement is invariant under any changes of sign of the a_j, we may and shall assume that all the a_j lie in the upper half-plane.

Any sum (1) is of the form

$$\sum_{j \in A} a_j - \sum_{j \in A'} a_j = \Sigma(A)$$

where A denotes the subset of all indices j for which $\varepsilon_j = +1$, and A' is the complement of A. Let F be the family of all subsets A of indices $\{1, 2, \ldots, n\}$, corresponding to sums $\sum(A)$ lying in the given unit circle. For $A, B \in F$ we have $|\Sigma(A) - \Sigma(B)| \leqq 2$; this is the only use that will be made of the assumption that the sums lie in a circle of unit radius.

We show that the family F meets the conditions of the preceding Lemma, where T is the subset of those indices j for which a_j lies in the first quadrant. For $A \supsetneqq B$, let ε_j and $\tilde{\varepsilon}_j$ be the sequences of ± 1's corresponding to A and B. For $j \in A - B$, we have $\varepsilon_j - \tilde{\varepsilon}_j = 2$; for $j \in B = A \cap B$, we have $\varepsilon_j - \tilde{\varepsilon}_j = 0$, and similarly for $j \in A' = A' \cap B'$ we have $\varepsilon_j - \tilde{\varepsilon}_j = 0$. Therefore

$$\begin{aligned}\Sigma(A) - \Sigma(B) &= \Big(\sum_{j \in A-B} + \sum_{j \in B} + \sum_{j \in A'}\Big)(\varepsilon_j - \tilde{\varepsilon}_j) a_j \\ &= 2 \sum_{j \in A-B} a_j .\end{aligned}$$

If in addition $A - B \subseteq T$, then every summand a_j appearing in the last sum is of absolute value greater than one, because the sum of a set of complex numbers in the first quadrant is also in the first quadrant. Hence, $\Sigma(A)$ and $\Sigma(B)$ cannot both lie in the same unit circle. Thus, if A and B belong to F, and $A \supseteqq B$, then $A = B$.

A similar argument on the second quadrant shows that the family F meets the second condition of the Lemma. The proof is therefore complete.

11. An Application to Integration

We conclude this Chapter with an application of the Marriage Theorem to a far-out field, namely the theory of almost periodic functions. The aim of this theory is to characterize those functions $f(x)$ of the real variable x which can be expanded into an infinite series of exponentials:

$$f(x) \sim \sum_{\lambda} c_{\lambda} e^{i\lambda x}$$

where λ ranges over a countable set of reals, and where the sum converges in a sense which need not concern us here. The remarkable property of almost periodic functions, first established by Bohr, is the existence of a *mean value*

$$M(f) = c_0 = \lim_{T\to\infty} \frac{1}{2T} \int_{-T}^{T} f(t)\mathrm{d}t,$$

which in turn can be used to compute the 'Fourier' coefficients c_{λ} by the formula

$$c_{\lambda} = \lim_{T\to\infty} \frac{1}{2T} \int_{-T}^{T} f(t)e^{-i\lambda t}\mathrm{d}t.$$

Our purpose is to establish the existence of a mean value as quickly as possible. The argument becomes more transparent if it is presented at its natural level of generality, namely, in the context of group theory, and we shall do so. The present approach shows that the mean of an almost periodic function can be constructed in a way which is very similar to the construction of the Riemann integral of a real function.

Let $f(x)$ be a complex function of the variable x ranging over a group G. We define an ε-partition $\pi = \pi(f, \varepsilon)$ of the function f, where ε is a small positive number, as a partition of the set G into a finite number of blocks $B_1, B_2, \ldots, B_n$, with the property that

$$\sup_{x,\, y \in B_i} |f(x) - f(y)| < \varepsilon, \qquad 1 \leqq i \leqq n.$$

Denoting as usual by cBd the set $\{cxd\colon x \in B\}$, namely, the *translate* of the set B, define the translate of a partition π, written $c\pi d$, as the partition of the group G whose blocks are the translates of the blocks B of the partition π.

An *almost-periodic function* on the group G is defined as a complex function f with the property that, for every small positive number ε, an ε-partition $\pi(f, \varepsilon)$ of G can be found, such that every translate $c\pi(f, \varepsilon)d$ is also an ε-partition of f.

On the basis of this definition, it is easy to verify that the sum, product, and uniform limit of almost periodic functions is again an almost -periodic function. It is also verified without difficulty that the present definition is equivalent to the better known Bochner definition of an almost periodic function as a function $f(x)$ for which the set of translates $\{f_a: a \in G\}$, where $f_a(x)$ is defined as $f(ax)$, is conditionally compact in the topology of uniform convergence.

The existence of a mean value for every almost periodic function can now be stated as follows:

Theorem 11.1. To every complex almost-periodic function f on a group G, one canas sociate a complex number $M(f)$, called the mean of f, with the following properties:

(a) $M(f_a) = M(f)$, in words, the mean is invariant under translation;

(b) The mean $M(f)$ is the unique limit point, uniform in the parameters c and d, ranging over G of the set

$$\frac{1}{n}\sum_{i=1}^{n} f(ca_i d)$$

where a_i (for $1 \leqq i \leqq n$) is an arbitrary finite subset of G.

Proof. We imitate as closely as possible the construction of the Riemann integral of a continuous function on the interval [0, 1]. To this end, recall that such an integral can be defined as the limit of the (directed) set of numbers

$$\frac{1}{n}\sum_{i=1}^{n} f(a_i)$$

associated with an ε-partition $\pi(f, \varepsilon)$ of [0, 1] whose blocks are intervals of length $1/n$, and where a_i is a point arbitrarily chosen in the i-th interval.

Guided by this analogy, we introduce a special class of ε-partitions on the group G, to be called *minimal* ε-partitions, relative to a fixed function f. These shall be those ε-partitions $\pi = \pi(f, \varepsilon)$, such that the partitions $c\pi(f, \varepsilon)d$ are also ε-partitions for all c and d in G, and with the additional property that they have a minimal number of blocks; that is, that they shall not be a proper refinement of any other 'invariant' ε-partition.

If $\pi = \pi(f, \varepsilon)$ is a minimal ε-partition, with blocks $B_1, B_2, \ldots, B_n$, and $b_i \in B_i$, let

$$M(\pi, \{b_i\}) = \frac{1}{n} \sum_{i=1}^{n} f(b_i). \tag{11.1}$$

It follows from the definition of an ε-partition that, if $c_i \in B_i$ is another set of representatives of the partition π, then

$$|M(\pi, \{c_i\}) - M(\pi, \{b_i\})| \leqq \varepsilon. \tag{11.2}$$

We now come to the crucial part of the proof, namely, the following statement:

(*) If π and σ are two minimal ε-partitions with the same number of blocks $B_1, B_2, \ldots, B_n$ and $C_1, C_2, \ldots, C_n$, a common set of representatives of π and σ can be found, namely, a set $\{b_i\}$ such that $b_i \in B_i \cap C_i$, after reindexing the C's if necessary.

Here is the proof. Construct a relation R between the B's and the C's, joining B_i to C_j whenever the intersection $B_i \cap C_j$ is not empty. Every matching of the relation R gives a common set of representatives for the two partitions. Thus, all we have to do is verify that the condition of the Marriage Theorem is satisfied.

Suppose that the blocks $B_1, B_2, \ldots, B_r$ are related to fewer than r blocks, say $C_1, C_2, \ldots, C_s$, where $s < r$. This would mean that

$$C_1 \cup C_2 \cup \ldots \cup C_s \supseteq B_1 \cup B_2 \cup \ldots \cup B_r = A,$$

and hence, replacing the first r blocks of π by the blocks $C_1 \cap A, C_2 \cap A, \ldots, C_s \cap A$, we would obtain an ε-partition of which π is a proper refinement, contradicting the assumption that the partition π is minimal. This concludes the proof of statement (*).

From (*) we infer at once that, for any elements $c, d \in G$, and for minimal ε-partition π,

$$|M(\pi, \{b_i\}) - M(c\pi d, \{c_i\})| \leqq 2\varepsilon, \tag{11.3}$$

where $\{c_i\}$ is an arbitrary set of representatives for the translate $c\pi d$. Observe that $c\pi d$ is also a minimal ε-partition.

Indeed, the partitions π and $c\pi d$ have a common set of representatives d_i, by (*). From the triangle inequality we obtain

$$|M(\pi, \{b_i\})-M(c\pi d, \{c_i\})| \leqq |M(\pi, \{b_i\})-M(\pi, \{d_i\})| + |M(c\pi d, d_i)-M(c\pi d, c_i)|$$

and by (11.2) each term on the right side is at most ε.

To complete the proof, we show that if π and σ are two arbitrary minimal ε-partitions, then

$$|M(\pi, \{b_i\})-M(\sigma, \{c_i\})| \leqq 4\varepsilon. \tag{11.4}$$

This obviously implies that the set of 'approximate means' $M(\pi, \{b_i\})$ converges as $\varepsilon \to 0$. Say that π has n blocks and σ has m blocks. By (11.3) we have

$$\left| \frac{1}{n}\sum_{i=1}^{n} f(b_i) - \frac{1}{n}\sum_{i=1}^{n} f(b_i c_j) \right| \leqq 2\varepsilon,$$

adding these inequalities for $j = 1, 2, \ldots, m$ and dividing by m, we obtain

$$\left| \frac{1}{n}\sum_{i=1}^{n} f(b_i) - \frac{1}{mn}\sum_{j=1}^{m}\sum_{i=1}^{n} f(b_i c_j) \right| \leqq 2\varepsilon;$$

similarly, interchanging the roles of m and n,

$$\left| \frac{1}{m}\sum_{j=1}^{m} f(c_j) - \frac{1}{nm}\sum_{i=1}^{n}\sum_{j=1}^{m} f(b_i c_j) \right| \leqq 2\varepsilon,$$

so that an application of the triangle inequality gives

$$\left| \frac{1}{n}\sum_{i=1}^{n} f(b_i) - \frac{1}{m}\sum_{j=1}^{m} f(c_j) \right| \leqq 4\varepsilon$$

which is (11.4), written out.

In conclusion, note that (11.3) implies that the 'approximate means' are invariant under translations, and this property is preserved in the limit. The proof is therefore complete.

When the group G is the real line, it can be verified without difficulty that the mean, as defined in the Theorem, coincides with the mean defined by the formula

$$M(f) = \lim_{T\to\infty} \frac{1}{2T}\int_{-T}^{T} f(t)\mathrm{d}t$$

Another application of the Marriage Theorem to a similar situation yields the existence of Haar measure in locally compact topological groups.

12. Notes and Problems

12.1. Ordered structures

Combinatorial theory is probably the most fertile field of applications of the techniques of ordered set theory, initiated by Dedekind and pursued through the late forties; we hope the present paper is proof enough of this statement. Ordered set theory, which was for many years developed with a view to applications in algebra, is instead indispensable in combinatorics, if only because an ordered structure is at times the only kind of structure one can find in combinatorial objects.

Just as there are several characterizations of Boolean algebras, the lattice of subspaces of a finite vector space has been similarly characterized, in a series of works that began with von Staudt in the last century and ended with Garrett Birkhoff. The analogy between the Boolean algebra of subsets of a set and the lattice of subspaces of a vector space is in fact one of the themes of contemporary mathematics: it can be observed for example in the logic of quantum mechanics as well as in number theory and in the theory of special functions, in the analogy between hypergeometric functions and Heine's q-hypergeometric functions.

One aspect of this analogy is the parallelism between binomial coefficients and Gaussian polynomials; this can be carried very far, using even topological notions (the 'right' notion of convergence for Gaussian polynomials is convergence in the q-adic norm; in this way, one obtains a q-analog of the Poisson distribution). Purely order-theoretically, one obtains a parallel between distributive and modular lattices, the latter being characterized by the identity $(a \vee b) \wedge c = (a \wedge c) \vee b$ whenever $b \geqq c$. The lattice of subspaces of a finite vector space is a complemented modular lattice. No analog is yet known for modular lattices of the structure theorem for distributive lattices (Theorem 1 of Section 5), although Dilworth has proved that Corollary 2 remains valid for all modular lattices. This is perhaps the deepest known combinatorial result on modular lattices.

One can also guess at a 'modular' analog of Theorem 1 of Section 5; it should be a statement about faithfully representing a modular lattice satisfying suitable restrictions as a sublattice of the lattice of subgroup of an Abelian group.

Pursuing the same analogy on another level, a 'desideratum' of long standing is the development of a 'modular logic' which stands in the same relation to modular lattices as classical logic does to distributive lattices. It has been claimed that the foundations of quantum mechanics depend on such a logic (Birkhoff, von Neumann, Mackey); but a suitable axiom system has never been given. The main difficulty seems to be the notion of a quantifier. This entire area is still shrouded in mystery, perhaps because the foundations of quantum mechanics are still in a superfluid state.

It is perhaps worth remarking that Theorem 1 of Section 5 yields the solution of the word problem for distributive lattices. A free distributive lattice D_n with n generators $x_1, x_2, \ldots, x_n$ can be shown to be finite, and the ordered set of join irreducibles to be isomorphic to the Boolean algebra of subsets of an n-set. Using Lemma 2 of Section 5, a canonical representation is obtained of every element of D_n; two elements of D_n are 'the same' whenever they have he same canonical representation as joins of join-irreducibles. The analogous problem for free modular lattices stands unsolved.

Perhaps the most interesting open combinatorial problem relating to distributive lattices was posed by Dedekind: compute the size of D_n. Presumably, the exponential generating function of these numbers should satisfy a functional equation, but very little is known on the subject (v. Birkhoff for further references and for some numerical values). Note that the equality stated in Corollary 2 of Section 5 does not come from a 'functorial' correspondence between the set of join-irreducibles and the set of meet-irreducibles: this is what makes the result interesting.

Several results of Sperner type, somewhat different from the ones developed in Section 9, are to be found in the scattered literature (de Bruijn, Lena Sharney); using the same method, it is not hard to derive similar results for the lattice of divisors of an integer n, or more generally for the lattices of subgroups of a finite Abelian group. Results of the Sperner type have found applications to number theory; see the papers of Erdös.

There are a number of works, due largely to Kleitman and to various Hungarians, regarding the maximum number in a family of sets, whose members are restricted in other ways, for example, the number of elements in the intersection or union of two sets in the family is restricted. Some of them are proved by eventually reducing the statement to Sperner's theorem (such is the case for the crucial Lemma in Section 10 due to Kleitman), but no systematic theory exists, and it would be interesting to have one.

We cannot close this all-too-brief survey of what is probably one of the central themes of combinatorial theory without mentioning the class of lattices which most often arises in combinatorial work, and which is presently a fertile field of research, namely, geometric lattices.

Geometric lattices are finite lattices which obey the chain conditions, so that a rank function, r, may be defined, satisfying, $r(0) = 0$, $r(a) = 1$ for every atom a and $r(a \vee b) + r(a \wedge b) \leqq r(a) + r(b)$ for all a, b in the lattice. It is also assumed that every element of a geometric lattice is the join of atoms (= elements of rank 1). For an arbitrary geometric lattice L the *Whitney numbers* (of the second kind) are defined to be $W_k(L)$, the number of elements of rank k in L. In the three classical examples of Section 2 these are the binomial coefficients, the Gaussian polynomials and the Stirling numbers of the second kind respectively. One may ask which properties of binomial coefficients are shared by the Whitney numbers of a geometric lattice. It is known for instance that the binomial coefficients, the Gaussian polynomials and the Stirling numbers are *unimodal*, that is, they increase with k to an essentially unique maximum, and then decrease. Is this true for all geometric lattices? Also, can a matching be found in the bipartite graph formed by the elements of rank k and $k+1$? If these two questions were to be answered in the affirmative, Sperner's theorem for arbitrary geometric lattices would follow.

Geometric lattices are interesting for two reasons: first, because they give a generalization of 'synthetic' projective geometry in a new and fruitful direction; second, because their arithmetic is closely related to the problem of coloring a graph. We hope to thoroughly discuss these connections elsewhere.

12.2. Muirhead's Inequality et al.

Very little work has been done to develop a systematic theory of inequalities for symmetric polynomials. Muirhead's inequality barely scratches the surface, and does not include, for example, the classical Newton inequalities between elementary symmetric functions, namely

$$p_{k-1}p_{k+1} \leqq p_k^2,$$

where

$$(x+a_1)(x+a_2)\ldots(x+a_n)$$
$$= x^n + \binom{n}{1} p_1 x^{n-1} + \binom{n}{2} p_2 x^{n-2} + \ldots + p_n.$$

In a vague sense, Newton's inequalities are quadratic, whereas Muirhead's inequality is linear.

I. Schur produced a 'linear' inequality which lies beyond the scope of Muirhead's inequality, namely

$$[\nu+2\delta, 0, 0, a_4, \ldots] - 2[\nu+\delta, \delta, 0, a_4, \ldots] + [\nu, \delta, \delta, a_4, \ldots] \geqq 0.$$

This inequality is probably one of an entire class which can presumably be obtained by deeper convexity arguments.

Ultimately we could ask for an analog of the Hilbert Basis Theorem for inequalities between symmetric polynomials. In other words, is there a finite set of inequalities from which all other inequalities between symmetric polynomials 'follow'? The difficulty lies in the interpretation of the word 'follow'. In the classical case, namely, in the study of invariant identities, rather than inequalities, the interpretation was achieved by introducing ideals, but in the case of inequalities nothing similar has yet been devised.

The proof we have given of Muirhead's inequality relies on the convexity of the exponential function. Furstenberg has shown that the exponential is optimal in the sense that replacing it by any other convex function does not give a better inequality.

There is, however, a vast literature concerned with the study and extension of the order relation '**a** is an average of **b**'; consult the work of Blackwell, Sherman, Meyer, and others. Ryff has proved the continuous analog of the Lemma which we used to establish the necessity in Muirhead's Theorem. His technique is similar to Douglas-Lindenstrauss and it leads to a continuous version of Muirhead's inequality (Ryff).

12.3. MATCHINGS

The origin of the marriage theorem goes back almost to the nineteenth century. It seems that the result was already known to Frobenius, but, as has often happened, its importance was not realized until the thirties, when it was independently rediscovered by Richard Rado and Philip Hall. Since then, there has been a small flood of papers extending, generalizing and understanding the theorem in all directions. For a thorough survey of the literature on the Marriage Theorem since its inception, we refer to the recent survey paper by Mirsky and Perfect. One of the outstanding problems on matchings is to estimate the number of matchings of a relation. All present estimates are exceedingly crude (the earliest work on the subject was done by Marshall Hall).

The possibility of extending the Marriage Theorem to several dimensions does not seem to have been explored. Thinking rather crudely, one might replace the matrix of zeros and ones by a 'tensor' taking values zero and one. A little-known result of De Bruijn seems to point the way in this direction: a parallelepiped of sides $k_1 \times k_2 \times \ldots \times k_n$ can be covered by 'bricks' of sides $d_1 \times d_2 \times \ldots \times d_n$ if and only if each of the d's divides one of the k, a different one for each d. This also gives a many-dimensional extension of the dimer problem.

It would be remiss to omit an explanation of the origin of the nickname 'marriage theorem'. A relation between a set G of girls and a set B of boys is obtained by relating a boy b to a girl g, whenever b and g are acquainted with each other. If the relation has a matching, then it will be possible to marry each of the girls to a boy with whom she is acquainted, an important event indeed.

A deep minimax theorem has been proved by Rado for distributive lattices. Given two distributive lattices L and M, and families $l_1, l_2, \ldots, l_n$ of elements of L, and $m_1, m_2, \ldots, m_n$ of elements of M. When can we find elements a_i of L such that $a_i \leqq l_i$ and the map $a_i \to m_i$ extends to a homomorphism of the lattice generated by the a's onto the lattice generated by the m's? Rado's answer is the following. Let $b(x_1, x_2, \ldots, x_n)$ be an arbitrary Boolean polynomial in the variables $x_1, x_2, \ldots, x_n$, that is, a polynomial involving the operations $\cup$ and $\cap$, which are assumed to satisfy the distributive law as well as the other lattice identities. A necessary and sufficient condition is that

$$r(b(b_1, b_2, \ldots, b_n)) \geqq r(b(m_1, m_2, \ldots, m_n),$$

where r denotes the rank function of either lattice.

In this form, the Theorem can probably be generalized to modular and even to geometric lattices. Such a result for geometric lattices would be very useful indeed.

Note that Rado's Theorem includes the Marriage Theorem as a special case.

Dilworth's theorem extends to locally finite ordered sets by a standard compactness argument; it does not, however, extend to arbitrary ordered sets. Is there a 'measure-theoretic' analog of Dilworth's theorem, in the same spirit as the Douglas-Lindenstrauss Theorem? In this connection, the work of Rado, extending the notion of majorization of averages which we have used in the proof of Muirhead's Inequality, should prove significant. It has been proved by Szpilrajn that every finite ordered set can be faithfully represented in an ordered set which is the product of chains (the product of two or more ordered sets is defined in the only conceivable way in which it can be defined). The minimum number of factors in such a product is called *dimension* of the ordered set (a concept due to Dushnik and Miller). Is there a mini-max characterization of the dimension? For distributive lattices, Dilworth's Coding Theorem gives the answer.

It would be interesting to consider partitions of an ordered set subject to other conditions on the blocks. Which 'properties' of the blocks lead to a 'minimax' statement? As a wild example, require that every block be an unrelated set. Then it is easy to see that the minimum number of blocks is equal to the maximum size of a chain (as H. Crapo pointed out to us).

Another possibility is to limit the size of a maximal unrelated incomparable set in each block, or, at the opposite end, to limit the size of a maximum chain in each block, as we do in Section 9. Results of this kind might point the way to a general understanding of combinatorial 'minimax Theorems'. At present, the only systematic theories in this direction are linear programming and the Ford-Fulkerson theory based on their max flow-min cut theorem; but the arguments relating combinatorial mini-max results to the mini-max

Theorem of linear programming are somewhat contrived. We venture to predict that some day the two theories will be merged into one attractive whole.

References

[1] Starred items refer to material treated in the text.

[2] Berge, C.: *Theorie des graphes et ses applications*, Paris, 1958. Translated as The Theory of Graphs and its applications, London 1962.

[3] *Birkhoff, G.: Lattice Theory, *Amer. Math. Soc.* Colloquium Publications, Vol. XXV, (1967).

[4] *Birkhoff, G.: Tres observaciones sobre el algebra lineal. *Univ. Nac. Tucuman, Rev. Ser.* A, 5 (1946), 147–151.

[5] Chapman, H. W.: A Note on the Elementary Theory of Groups of Finite Order. *Messenger Math.* 42 (1913), 132–134.

[6] Cohn, P. M.: *Universal Algebra*, Harper and Row, 1965.

[7] Dantzig, G. B. and Hoffman, A. J.: *Dilworth's Theorem on Partially Ordered Sets* [63], 207–214.

[8] Dantzig, G. B. and Hoffman, A. J.: On a Theorem of Dilworth, Contributions to Linear Inequalities and Related Topics, *Annals of Mathematics Studies*, No. 38.

[9] Dedekind, R.: Uber Zerlegungen von Zahlen durch ihre grosstengemeinsamen Teiler, *Festschrift Techn. Hoch.* Braunschweig (1897), and *Ges. Werke*, Vol. 2, 103–148; 2 Uber die von drei Moduln erzeugte Dualgruppe, *Math. Ann.* 53 (1900), 371–403, and *Ges. Werke*, Vol. 2, 236–71.

[10] *Dilworth, R. G.: A Decomposition Theorem for Partially Ordered Sets, *Annals of Mathematics* 51, 1 January 1950 (161–166).

[11] Dilworth, R. G.: Structure and Decomposition Theory of Lattices. *Am. Math. Soc.*, Proceedings of Symposia in Pure Mathematics, Vol. II, (1960), 3–16.

[12] *Douglas, R. G.: On Extremal Measures and Subspace Density. *Michigan Mathematical Journal* 11 (1964) 243–246.

[13] *Dushnik, B. and Miller, E. W.: Partially Ordered Sets, *Amer. J. Math.* 63, (1941) 600–610.

[14] Erdös, P.: On A Lemma of Littlewood and Offord. *Bull. Amer. Math. Soc.* 51, 898-902 (1945).

[15] Ford, L. R. and Fulkerson, D. R.: *Flows in Networks*. Van Nostrand, Princeton, 1962.

[16] Frobenius, G.: Uber Matrizen aus nicht negativen Elementen. *Sitzb. Preuss. Akad. Wiss.* 1912, 456–477.

[17] Frobenius, G.: Uber zerlegbare Determinanten. *Sitzb. Preuss. Akad, Wiss.* 1917, 274–277.

[18] Fulkerson, D. R.: Note on Dilworth's Decomposition ofr Partially Ordered Sets, *Proc. Amer. Math. Soc.* Vol. 7 (1956) pp. 701–702.

[19] Gallai, T. and Milgram, A. N.: Verallgemeinerung eines graphentheoretischen Satzes von Redei. *Acta. Sci. Math.* (Szeged) 21 (1960), 181–186.

[20] Hall, M. Jr.: An Existence Theorem for Latin Squares. *Bull. Am. Math. Soc.* 51 (1945), 387–388.

[21] Hall, M. Jr.: Distinct representatives of subsets. *Bull. Am. Math. Soc.* 54 (1948) 922–926.

[22] Hall, M. Jr.: An Algorithm for Distinct Representatives. *Am. Math. Mo.* 63 (1956), 716–717.

[23] Hall, M. Jr.: *A Survey of Combinatorial Analysis.* Surveys in Applied Mathematics. IV. Some Aspects of Analysis and Probability, pp. 37–104, New York 1958.

[24] *Hall, P.: On Representatives of Subsets, *J. London Math. Soc.* vol. 10 (1935) pp. 26–30.

[25] *Halmos, P. and Vaughan, H. E.: The Marriage Problem. *Am. J. Math.* 72 (1950) 214–215.

[26] *Hardy, G. H., Littlewood, J. E. and Polya, G.: *Inequalities.* Cambridge University Press, 1934.

[27] Higgins, P. J.: Disjoint Transversals of Subsets. *Canad. J. Math.* 11 (1959), 280–285.

[28] Hoffman, A. J.: *Some Recent Applications of the Theory of Linear Inequalities to Extremal Combinatorial Analysis* [2], 113–127.

[29] Hoffman, A. J.: Linear Programming. *Appl. Mech. Rev.* 9 (1956), 185–187.

[30] Hoffman, A. J., and Kuhn, H. W.: *On Systems of Distinct Representatives* [63] (1956), 199–206.

[31] Hoffman, A. J., and Kuhn, H. W.: Systems of Distinct Representatives and Linear Programming. *Am. Math. Monthly* 63 (1956), 455–460.

[32] Isbell, J. R.: Birkhoff's Problem 111. *Proc. Am. Math. Soc.* 6 (1955), 217–218.

[33] Kellerer, H. G.: Allgemeine Systeme von Repräsentanten. *Z. Wahrscheinlichkeitstheorie* 2 (1963–1964), 306–309.

[34] *Kleitman, D.: On a Lemma of Littlewood and Offord on the Distribution of Certain Sums. *Math. Zeitschr.* 90, (1965) 251–259.

[35] Kleitman, D.: Families of Non-disjoint Subsets, *JCT* 1, 1 (1966) 153–155.

[36] Kleitman, D.: On Subsets Containing a Family of Non-Commensurable Subsets of a Finite Set, *JCT* 1, 2 (1966) 297–299.

[37] *Kleitman, D. J., and Rota, G.-C.: *Some extremal results of Sperner Type* (to appear).

[38] König, D.: Uber Graphen und ihre Anwendung auf Determinantentheorie und Mengenlehre. *Math. Ann.* 77 (1916), 453–465.

[39] König, D.: *Theorie der endlichen und unendlichen Graphen.* Leipzig, 1936.

[40] König, D.: *Theorie der Graphen,* New York, Chelsea, 1950.

[41] Kreweras, G.: Extension d'un theorème sur les repartitions en classes. *Compt. Rend. Acad. Sci.* 222 (1946) 431–432.

[42] Kuhn, H. W. and Tucker, A. W.: Linear Inequalitites and Related Systems. *Annals of Mathematics Studies* No. 38, Princeton Univ. Press, 1956.

[43] *Lindenstrauss, Joram: A Remark on Extreme Doubly Stochastic Measures *American Mathematical Monthly* 72, 4 (1965) 379–382.

[44] *Littlewood, J. E. and Offord, C.: On the number of real roots of a random Algebraic Equation (III). *Mat. Sbornik* (*Rec. Math.*) N.S. 12, 277–285.

[45] *Lubell, D.: A Short Proof of Sperner's Lemma. *JCT* 1, 2 (1966) 299.

[46] * Maak, Wilhelm: *Eine neue Definition der fastperiodischen Funktionen.* Hamburg, Mathematisches Seminar (1935).

[47] Maak, Wilhelm: Eine neue Definition der fastperiodischen Funktionen. *Abh. Math. Sem. Hamburg. Univ.* 11 (1936), 240–244.

[48] Maak, Wilhelm: Ein Problem der Kombinatorik in seiner Formulierung von H. Weyl. *Math.-Phys. Semesterber.* 2 (1952), 251–256.

[49] Mann, H. B. and Ryser, H. J.: Systems of distinct representatives. *Am. Math. Monthly* 60 (1953), 397–401.

[50] Marcus, M. and Minc. H.: Some Results on Doubly Stochastic Matrices. *Proc. Am. Math. Soc.* 13 (1962), 571–579.
[51] Mendelsohn, N. S. and Dulmage, A. L.: Some generalizations of the Problem of Distinct Representatives. *Canad. J. Math.* 10 (1958), 230–241.
[52] Menger, K.: Zur allgemeinen Kurventheorie. *Fund. Math.* 10 (1927), 96–115.
[53] Mirsky, L.: Results and Problems in the Theory of Doubly-Stochastic Matrices. *Z. Wahrscheinlichkeitstheorie* 1 (1962–1963), 319–234.
[54] *Mirsky, L. and Perfect, Hazel: Systems of Representatives, *J. Mathematical Analysis and Applications*, Vol. 15, No. 3 'September 1966), 520–568.
[55] *Muirhead, R.: Some Methods Applicable to Identities and Inequalities of Symmetric Algebraic Functions of *n* Letters. *Proc. Edin. Math. Soc.*, 21 144–157 (1903).
[56] Ore, O.: Graphs and Matching Theorems. *Duke Math.* J. 22 (1955) 625–639,
[57] Ore, O.: Theory of Graphs. *Am. Math. Soc. Coll. Publ.* Vol. 38, Providence. 1962.
[58] Perfect, Hazel: Symmetrized Form of P. Hall's Theorem on Distinct Representatives, *Quart. J. Math. Oxford* (2) (1966) 303–306.
[59] *Perles, M. A.: A Proof of Dilworth's Decomposition Theorem for Partially Ordered Sets. *Israel J. Math.* 1 (1963), 105–107.
[60] *Rado, R.: Bemerkungen zur Kombinatorik im Anschluss an Untersuchungen von Herrn D. Konig, *Berliner Sitzungsberichte*, Band 32 (1933), S. 68.
[61] Rado, R.: A Theorem on General Measure Functions. *Proc. London. Math. Soc.* (2), 44 (1938), 61–91.
[62] Rota, G.-C.: On the Foundations of Combinatorial Theory. I. Theory of Möbius Functions, *Zeits, f. Wahrscheinlichkeitsrechnung* 2 (1964), 340–363.
[63] Rota, G.-C.: The number of Partitions of a Set, *Am. Math. Monthly*, 71, 5 (1964) 499–504.
[64] *Rota, G.-C.: *Dilworth Numbers and Partitions of an Ordered Set* (to appear).
[65] Ryser, H. J.: A Combinatorial Theorem with an Application to Latin Rectangles. *Proc. Am. Math. Soc.* 2 (1951), 550–552.
[66] Ryser, H. J.: *Combinatorial Mathematics.* Carus Mathematical Monographs. No. 14, 1963.
[67] Scorza, G.: A Proposito di un teorema del Chapman. *Boll. Unione Mat. Ital.* 6 (1927), 1–6.
[68] Shannon, C. E.: Prediction and Entropy of Printed English, *Bell System Tech. J.* 30, 50–64 (1951).
[69] Sherman, S.: On A Theorem of Hardy, Littlewood, Polya, and Blackwell. *Proc. of the Nat. Ac. of Sc.*, 37, 12 (1951) 826–831.
[70] Shohat, J. A. and Tamarkin, J. D.: The Problem of Moments. *Amer. Math. Soc. Mathematical Surveys* Vol. 2, New York, 1943.
[71] *Sperner, E.: Note zu der Arbeit von Herrn B. L. van der Waerden: Ein Satz uber Klasseneinteilungen von endlichen Mengen. *Abh. Math. Sem. Hamburg. Univ.* 5 (1927), 232.
[72] van der Waerden, B. L.: Ein Satz uber Klasseneinteilungen von endlichen Mengen. *Abh. math. Sem. Hamburg. Univ.* 5 (1927), 185–188.
[73] *Whitney, H.: The Abstract Properties of Linear Dependence, *Amer. J.* 57 (1935), 507–533.